WINDS
WAVES &
WARRIORS

WINDS
BATTLING THE SURF AT
WAVES &
NORMANDY | TARAWA | INCHON
WARRIORS

Thomas M. Mitchell

LOUISIANA STATE UNIVERSITY PRESS BATON ROUGE

Published by Louisiana State University Press

Manufactured in the United States of America
First printing

DESIGNER: Michelle A. Neustrom
TYPEFACES: Sina Nova, text; Futura BT, display
PRINTER AND BINDER: Sheridan Books, Inc.

Maps, illustrations, and diagrams created
and adapted by Mary Lee Eggart

Cataloging-in-Publication Data are available at the Library of Congress.
ISBN 978-0-8071-7223-0 (cloth: alk. paper)

The paper in this book meets the guidelines for permanence and durability
of the Committee on Production Guidelines for Book Longevity
of the Council on Library Resources. ♾

This book is dedicated to Robert O. Reid,
professor emeritus at Texas A&M University,
teacher, mentor, and friend.

CONTENTS

ACKNOWLEDGMENTS

I WISH TO EXPRESS MY DEEPEST GRATITUDE to the many friends, colleagues, and family who helped make this book possible.

Charles Bates and John Crowell, who played key roles in preparing the forecasts of beach and surf conditions for the Normandy beaches on D-Day not only provided invaluable information by means of email exchanges and telephone conversations, but each spent an entire afternoon with me in extended personal interviews. I cannot begin to express my gratitude to these two fine gentlemen for the generous gifts of their time and for their contributions to the success of the D-Day landings.

Nor can I ever adequately thank Dr. Walter Munk for the afternoon and evening he spent telling me about the Ocean Forecasting course he and Dr. Sverdrup taught at the Scripps Institution of Oceanography. They taught several hundred army, navy, and marine officers the basics of physical oceanography and forecasting ocean and surf conditions, and saved the lives of thousands of these brave warriors who stormed enemy-held beaches. Charles Bates and John Crowell were among their students, as was Robert Reid.

I have dedicated this book to Professor Robert O. Reid, my major professor and mentor at Texas A&M University, but I want to acknowledge him here also. Professor Reid took me on as an aspiring graduate student whose main knowledge of the ocean was as a sport fisherman and boater. He molded me into a serious physical oceanographer who still stands in awe of his mentor. Bob Reid was a quiet-spoken man of infinite patience. He held weekly, two-hour tutoring sessions with each of his new graduate students. When he first told me of this policy and we looked at our schedules, the only time we had mutually available was 2:00–4:00 pm on Friday afternoon. What an awful time! I wanted to start my weekend by then, but after only one or two sessions these two hours became the highlight of my week. It is no exaggeration to say that Bob Reid gave me my career as an oceanographer.

I wish also to recognize that wonderful institution known as the Bread Loaf Writers' Conference, the oldest writers' conference in the US. I attended thirteen ten-day summer sessions at the old Bread Loaf Inn under the leadership of Michael Collier and his able assistant Noreen Cargill. It was there that I gained the knowledge and confidence to attempt to write a book. I was fortunate to be able to work with such outstanding nonfiction workshop leaders as Patricia Hampl, Bill Kittredge, Ted Conover, David Shields, Tom Bissell, Jane Brox, and Ted Genoways.

Another Bread Loafer I wish to recognize is David Haward Bain, professor at Middlebury College, whom I met and befriended during my first summer on the mountain. David believed in this project and always offered words of encouragement to keep me going.

I can never thank my editor at the LSU Press, Margaret Lovecraft, enough for sponsoring this book. We were introduced by Kris Elmore (director of development for the LSU College of Engineering, who deserves my thanks for that introduction) and after a lengthy conversation, Margaret invited me to send her a proposal for *Winds, Waves, and Warriors*. The proposal was accepted for publication by the Press and Margaret worked with me to develop and polish the manuscript. Margaret's guidance and patience as I learned the intricacies of publishing a book were immeasurable and were all that got me through that process. Many, many thanks to you, Margaret.

Dr. George Z. Forristall, a colleague with whom I worked many years in the oil industry, read some early excerpts of the oceanographic concepts and commented on them for correctness and readability. Then he read a late draft of the entire manuscript and commented on it, as well. Any omissions or errors are the responsibility of the author alone, not Dr. Forristall.

Many thanks are due to Joris de Vroom, consultant for Metocean Modeling at Infoplaza Marine Weather in the Netherlands, who prepared the hindcast time series plots of tide heights at Normandy.

ACKNOWELDGMENTS

I want to thank my writing critique group, the Talespinners—Randall Best, Nelleva Newton, and Karen Swensson—for reading and rereading many drafts and versions of this manuscript. It would not have been the same without their suggestions, help, and guidance.

I owe a huge debt of gratitude, which can never be fully paid to my wife, Dottie, for her never-ending support and encouragement. When I first told her I wanted to write a book about WWII I'm sure she had no idea how much she would be involved in it when she said to go ahead. But as it slowly took shape over several years and she was asked to read yet another chapter, she did so willingly and always had useful comments. Dottie supported me throughout this endeavor not only with her editorial prowess but more importantly, with her love. Thank you forever, Dottie.

WINDS
WAVES &
WARRIORS

1

PRELUDE TO NORMANDY

INVADERS HAVE USED THE SEAS to reach their enemies' shores for as long as warriors have paddled canoes or sailed ships. Sometimes the ocean allowed them to slip up to their adversary's coast and disembark on his beach. Other times it destroyed the attacking forces while they were still at sea. At first, luck, or the gods, seemed to decide which it would be, but as scientists studied the ocean and learned more about it, military planners used this knowledge to recognize and take advantage of opportunities the sea offered and to foresee and avoid dangers that it posed.

US sailors and ships have fought many battles with fierce storms throughout the nation's military history. These battles have been well chronicled, but far less well known are the battles foot soldiers of the army and Marine Corps fought going ashore. During World War II, countless soldiers and marines lost their lives at the ocean's edge in what oceanographers call the near shore, or surf zone—roughly the area extending from the beach seaward to where waves begin to break. The ocean here can be anywhere from flat calm to unbelievably violent. When the sea is on a rampage, it is no place for humans or machines.

Military tacticians must consider many aspects of coastal conditions when they decide to invade a hostile land from the sea. Winds, waves, currents, tides, and tricky beach and bottom conditions have all confronted soldiers and marines whose only goal in life during an amphibious landing was dodging the next bullet or the next wave.

Wind is the usual starting point for discussions of physical oceanographic phenomena because it drives most of the others. Strong winds over the sea have long been observed to be accompanied by high waves, and people deduced (correctly) that the winds caused the waves. Winds and waves presented problems for the allies on D-Day at Normandy, but they were less troublesome that day than they would have been the previous day or the day after.

There are several causes of ocean currents. Wind-driven currents and those caused by tides are the two most important to amphibious operations. Currents during landings at Normandy, Sicily, and some of the Pacific islands pushed landing craft far away from their intended landing sites, resulting in confusion and loss of personnel and equipment.

Tides can cause the water depth to vary by as much as fifty feet over a six-hour period. Tides are different from place to place, and from day to day at the same place, and can present opportunities or enormous difficulties when planning and conducting amphibious landings. At Normandy, General Dwight Eisenhower surprised the Germans by landing on a low tide instead of a high tide as they expected.

Soldiers and marines had to contend with bottom conditions ranging from coral reefs that stopped landing craft hundreds of yards offshore at Tarawa Atoll in the Pacific to soft mud bottoms that would not support a man's weight at Inchon, Korea. The reef and tides contributed to the excessive casualties suffered by marines at Tarawa. With very intricate planning, General Douglas MacArthur used the unfavorable bottom and tidal conditions at Inchon to surprise the North Koreans, who thought no one would dare attack at a location with so many natural obstacles.

When people have worked in harmony with the sea as General MacArthur did at Inchon, they usually have been successful. One needs only to see photographs of the devastation left in the wake of a major hurricane to realize what an angry ocean can do to people, structures, and machines that are in the wrong place at the wrong time. And so it is with military operations.

US troops have landed both when the ocean contributed to the success of a mission and when it presented great difficulties. What made the difference? Usually, it was the thoroughness with which the planners considered the effects the ocean and weather could have on the operation and how well they addressed their potential for disruption.

With the possible exception of the Normandy beaches on D-Day, information on enemy beaches was never complete. Sometimes planners attempted to get the proper information only to have something go wrong with the information gathering or to have the information interpreted incorrectly. Sometimes amphibious landings went awry because planning only considered the ocean as something to float the boats that moved the troops to shore. The goal for the staff planning an amphibious operation and for those carrying it out is to get the landing craft through the surf to the beach in a place where the soldier can get ashore without getting killed either by the sea or by the enemy.

~

GENERAL WINFIELD SCOTT faced such a dilemma in 1847 as he stood on the deck of his flagship, the *Massachusetts,* staring shoreward at the city of Vera Cruz, Mexico. He was following orders from President James Polk to invade the country at this coastal town, then march inland three hundred miles and capture the capital, Mexico City. This operation would soon become America's first major amphibious landing.

General Scott would have preferred to disembark his troops at a proper dock, but the rivers along this part of the Mexican coast were too small and shallow for the large American ships to enter. The answer was to land the soldiers on the beach, a strategy that later became known as an amphibious landing. To do this, he needed special boats that did not exist in the inventories of either the army or navy. Scott specified the design of a new type of boat he called "surfboats,"[1] which were built in Philadelphia for the army Quartermaster's Department.

General Scott recognized that he needed a landing beach with a firm sand bottom protected from heavy surf. He was not familiar with the beaches near Vera Cruz, but he knew that Commodore David E. Conner, commander of the US Navy's Home Squadron, had been blockading the port of Vera Cruz for many months. Scott sailed to Vera Cruz and conferred with Commodore Conner, who recommended Collado Beach, about two miles south of Vera Cruz, as the landing site (see Map 1).

Collado Beach was wide and sandy, with enough room for all the surfboats to land and discharge their troops. The bottom sloped sufficiently for the boats to get close to shore before grounding, and it was out of range of the large guns at the fort. It was protected from waves on all sides except for a narrow passage on the southeast that was large enough for the American ships to enter. Swell

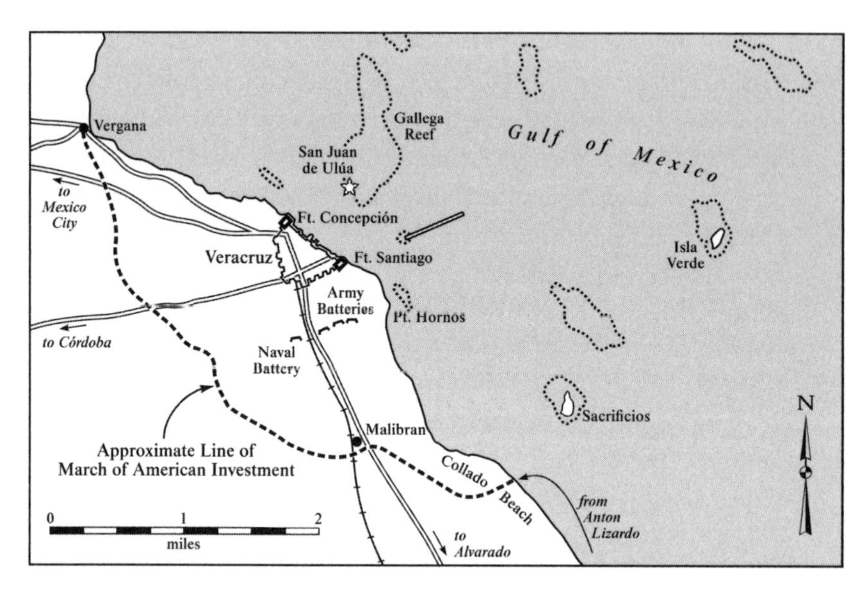

MAP 1

Vera Cruz and vicinity. General Winfield Scott's fleet sailed from their rendezvous/training area at the small town of Anton Lizardo to the anchorage behind Sacrificios Island to stage for their landing at Collado Beach on the morning of March 9, 1847.

waves from distant storms could have wrecked the surfboats at a more exposed site. The bottom material was firm sand. The tide range at Collado Beach was less than one meter, which simplified planning since the landing did not have to take place during a narrow time period at some specific tide stage. From what the Americans could see from offshore, the Mexicans had made no defensive preparations at Collado—an ideal situation for an amphibious landing.

The plan was not without complications, though. The anchorage at Collado was well protected, but it was too small to accommodate Scott's many troop transport ships and the naval gunships. Commodore Conner conceived a plan to solve this problem.[2] If Scott agreed to transfer his troops to the much larger navy ships, fewer ships would need to enter the small harbor. It frustrated General Scott to turn the maneuver into a joint operation and to have to share the glory with the navy, but he was a practical man and agreed.

Orders to prepare for the landing were issued, and preparations began the following morning, March 8. However, increasing winds and cloudiness warned of an approaching "norther" that might prevent the landing. Northers are severe storms that occur in the winter and spring with gale-force north winds up to fifty knots.[3] They originate in the arctic regions of North America and sweep southward across Canada, the central United States, and into the Gulf of Mexico. The mass of dense, bitterly cold, dry air is preceded by a cold front with strong north to northwest winds. Low clouds along the front are wet and stormy and often appear bluer than the sky above them. This has given rise to the colloquial name "blue norther" for these storms. When the storm's cold air flows over the warm, open waters of the Gulf, it becomes unstable and produces strong, gusty winds, which in turn can create waves up to twenty-five feet high. Northers are second only to hurricanes and tropical storms in the Gulf of Mexico in their ferocity and destructive power. Although they normally last only a day or two, three at most, Scott knew the one they expected spelled disaster if he did not avoid it. He and Commodore Conner canceled the operation and rescheduled it for the next day.

The storm did not materialize, and March 9, 1847, dawned with clear blue skies and a gentle breeze from the southeast. The operation could proceed. The landing parties headed for shore, expecting to fight their way across the water to the beach, but the Mexicans did not fire a single shot. Only ten minutes after the signal to attack had been given, 2,595 men of the first wave landed without a mishap.[4] The surfboats returned to the troop ships for the second and third waves of soldiers. By approximately 2200 hours, nearly eight thousand men had landed on the beach at Collado. General Scott's well-planned and well-executed amphibious landing was a brilliant success and broke the stalemate in the land war.

Despite Scott's initial good fortunes, a series of northers began on March 12 and lasted until March 17, severely hampering the landing of equipment.[5] Several ships broke from their moorings and were wrecked. Many cavalry horses drowned, and much equipment was lost. Small boats and surfboats were scattered for miles along the beach. The soldiers on shore were cut off from their supplies and support from the ships. The weather finally broke on March 17, and the landing of supplies resumed. General José Juan Landero, commander of San Juan de Ulua, surrendered Vera Cruz to General Scott on March 29, 1847. Scott then led his army three hundred miles inland and captured Mexico City, ending the war with Mexico.

OVER THE NEXT SEVENTY YEARS, American amphibious warfare doctrine changed very little. Although the US Army made several amphibious landings during the Civil War, it used the same tactics General Scott used at Vera Cruz. During the Gallipoli Campaign of World War I, British and Commonwealth soldiers conducted several amphibious landings in attempts to dislodge the Ottoman Turks from forts they occupied along the Dardanelles. These amphibious attacks of the Gallipoli Campaign were military failures for the British; however, they were studied in detail by the militaries of many nations, including the United States, Great Britain, and Japan, to improve and modernize their amphibious tactics and equipment.

US military planners realized that the next major conflict would likely be fought in the Pacific with Japan as the major adversary, and the navy would be called upon to defend the Philippines and other US interests there.[6] The navy and Marine Corps at that time were operating under War Plan Orange, a series of oft-updated plans for the protection of the Philippines and western Pacific nations from Japanese attack. Major General John A. Lejeune, Commandant of the Marine Corps, realized that War Plan Orange had no specific plan for amphibious operations, which he considered a serious omission. He assigned marine Lieutenant Colonel Earl H. Ellis to develop the requirements of a plan for amphibious operations to counter the Japanese aggression that was expected somewhere among the thousands of islands in the western Pacific. Ellis had been a faculty member at the Naval War College and was an ideal author for the study. His work culminated in Operation Plan 712, *Advanced Base Operations in Micronesia,* and was endorsed by General Lejeune July 23, 1921.[7] Twenty years later when the United States retaliated for the attack at Pearl Harbor, the navy's strategy of island-hopping across the Pacific was based on Ellis's document.

Ellis reasoned that the best strategy for capturing enemy-held islands would be to approach each island from the sea and sail through an opening in the surrounding coral reef to land directly on the island's beach. He included general information on ocean currents and advised that the currents were "much influenced by the winds" and were stronger around the islands. He recognized that the tide range in the western Pacific was not extreme but that tidal currents were "generally very strong and cannot be calculated upon to turn with high and low water." He warned of the hazards of navigating in these waters.[8] Implicit in Ellis's work was the admonition to know the ocean in the location where an amphibious operation was being planned. His advice was general and could not be site-specific because the islands are spread over a vast oceanic area where conditions vary immensely, and no one had any idea where a conflict with the Japanese might take place.

The Marine Corps refined its amphibious doctrine with the publication in 1931 of the detailed textbook *Tentative Manual for Landing Operations.*[9] It was integrated into the Corps' courses taught at Marine Corps Base Quantico, Virginia, in 1934. As World War II drew nearer, this "tentative" manual was tested in naval exercises and was adopted as doctrine by the marines, as well as by the army and navy by the time the Japanese attacked Pearl Harbor.

With the Marine Corps' emphasis on amphibious operations, the navy knew it would need ships and landing craft to carry the troops and their equipment to shore. They started in-house programs to design the landing craft, personnel (LCP) and the landing craft, vehicle, personnel (LCVP), but the marines insisted that a boat they called the Higgins boat (named for its designer, Andrew Jackson Higgins) was far superior. During the 1930s, Andrew Higgins built shallow draft boats in his New Orleans facility for use by fur trappers and fishermen in the swamps and marshes of Louisiana. The trappers needed a rugged flat-bottom boat that could be run aground without damage to check their traps. Over a period of several years, Higgins developed his Eureka boat to fill the needs of these trappers. The Eureka was so rugged that it could run over obstacles such as floating logs without damage. Oil companies found the Eureka boats very useful when they started exploring for oil in the marshes and nearshore waters of the Gulf of Mexico. In the late 1930s, Eureka boats and other Higgins boats were bought by various oil companies and exported to Venezuela, Colombia, Peru, Mexico, and several other South American countries.[10] Many of these latter countries bought the Eurekas for use as gunboats. The US Army Corps of Engineers and the Biological Survey Agency used Higgins boats for their shallow-water work.

Andrew Higgins had tried to interest the navy in his Eureka boat as a shallow-water landing craft as early as 1928 but initially was unsuccessful. The navy politely informed him that it was designing its own landing craft.[11] The navy's Bureau of Construction and Repair held sea trials of five existing-design fishing boats from New England boat companies in the summer of 1936 to select a troop-landing craft. After the trials, the selection board concluded that none of the New England fishing boats was satisfactory for the task at hand. Higgins had not been invited to participate, but he persisted and by February 1939 was allowed to enter one of his Eureka boats in Flex 5, the navy's annual full scale war games and at-sea experiments to test new equipment and tactical concepts. The Eureka so impressed the senior member of the selection board that he declared it superior to the other boats tested, but bureaucratic patronage resulted in the navy's design being selected. Disappointed but undeterred, the New Orleans boat builder persisted, was invited to participate in several direct competitions, all of which his Eurekas easily won, and finally, on November 18, 1940, was awarded a contract to build 335 thirty-six-foot Eurekas as the navy's standard LCP. According to marine General Holland Smith, "Andrew Higgins, a fighting Irishman, won the opening phase of the boat battle singlehanded, with loud Marine applause."[12]

As pleased as General Smith was with Higgins's LCP, he knew that when his marines landed on an enemy-held shore they needed the support of artillery and tanks. He turned to Higgins to design lighters that could offload tanks and artillery pieces from transport ships and land them on enemy beaches. The lighters would need to be landing craft capable of approaching a beach through the surf and unloading tanks and artillery onto the beach. Higgins was shown a photograph of a Japanese landing craft that had a ramp built into its bow. The ramp could be lowered and raised. Andrew Higgins immediately saw the improvement this would make to his Eureka LCP and also how it would form the basis for a tank lighter. The LCP with the ramp bow and some other improvements suggested by the marines became the LCVP, also known as the Higgins boat, and was the workhorse personnel landing craft of World War II. The Higgins boats were produced by the thousands.[13] The LCVP's bow ramp protected its occupants as it approached shore and then could be lowered so soldiers and marines could disembark quickly on an enemy-held beach. This ramp was the main advantage of the Higgins design.

The landing ship, tank (LST), and the landing ship, dock (LSD), were developed in 1942 and 1943, respectively, to deliver tanks, trucks, artillery, and other large vehicles to a war zone.[14] The bow of the LST had barnlike doors that swung outward to open virtually the full width of the main cargo hold. The LST had a flat bottom and could run into very shallow water to unload vehicles through its doors. When the ship reached the beach and grounded, the doors were opened, a ramp was lowered, and trucks, jeeps, artillery, or other heavy cargo were simply driven down the ramp onto the shore. LSTs were used in this fashion in the landings at Normandy, Bougainville, Salerno, Sicily, New Guinea, Saipan, Inchon, and elsewhere in Europe and the Pacific.[15]

The LSDs were even larger than LSTs and were used to carry cargoes of heavy vehicles and other amphibious landing craft into battle zones. Tanks, artillery pieces, etc. were loaded in amphibious landing craft, then were loaded into what was called the well deck of the LSD. When the LSD reached the battle area, the well deck was flooded and then the smaller amphibious landing craft were launched with their cargoes of tanks, etc. for the short trip to the beach.

Also at this time, marine and army aviation doctrine and aircraft were developed to support the soldiers and marines storming an enemy beach.

Soon after the United States entered World War II, military planners at the army's Washington, DC, Weather Central realized that when soldiers and marines jumped off their landing craft, they might leap into ocean conditions as deadly as the German or Japanese gunfire they faced. They sought and obtained authorization and funding to begin graduate-level courses in meteorology at five of the nation's top meteorology schools to teach weather forecasting to military personnel of all the services.[16] Shortly thereafter, the Weather Central officers identified the additional requirement for weather officers who could prepare site-specific forecasts of beach conditions at potential amphibious landing sites. An advanced course was started at Scripps Institution of Oceanography to teach ocean forecasting to selected graduates of the meteorology course.

By early 1944, the allies had assembled vast numbers of soldiers, airmen, sailors, ships, airplanes, and every other machine of war throughout the United Kingdom in preparation for Operation Overlord, the greatest amphibious attack the world has ever witnessed. Parallels between General Scott's landing at Vera Cruz and General Eisenhower's landing at Normandy ninety-seven years later are remarkable. Military experience is no exception to the adage that history has a way of repeating itself.

Both landings were postponed one day due to forecasts of adverse weather. The forecast for Normandy was accurate, and disaster would have ensued had General Eisenhower ignored the warnings. At Vera Cruz the forecast caused

General Scott to delay the landing even though a strong norther did not actually occur.

Proper planning for the weather and ocean conditions at the landing sites made both landings successful, but severe storms later delayed the postinvasion resupply. As a result, both operations lost many boats and much equipment in the storms. The men on shore were cut off from their supplies and food until the storms abated.

American forces during both invasions had nearly complete mastery of the enemy's primary means to counterattack the invaders. At Vera Cruz, Commodore Conner controlled the sea and prevented any seaborne relief of the beleaguered fort. During the Normandy landing, the Allied air forces controlled the skies, making the German Luftwaffe ineffective.

Generals Scott and Eisenhower both appreciated the need for reconnaissance of the landing beaches. Scott sought Commodore Conner's advice about beach conditions, then personally reconnoitered Collado Beach with members of his staff before making his final decision. General Eisenhower had a large staff of experts in London studying every facet of the operation for months before launching Operation Overlord.

The US Navy played an indispensable role in both operations. It bombarded suspected enemy positions with high-explosive shells before the soldiers started ashore and manned the boats that carried the soldiers to the beach.

Hundreds of ships participated in the Vera Cruz invasion and thousands at Normandy. Both invasions used landing craft designed expressly to transfer men and equipment ashore. At Vera Cruz, General Scott had his surfboats. At Normandy, Allied forces used landing craft of several different designs and sizes to transfer soldiers and equipment to the beaches.

There was one great difference, however: the Mexicans did not oppose the Vera Cruz landing, while the German Army fought fiercely at Normandy. The casualty lists for the two invasions were vastly different.

2

TRAINING OCEANOGRAPHIC METEOROLOGISTS

ENGAGED IN A WAR ON TWO CONTINENTS, the United States in 1942 began mobilizing men and materiel. As the requirements were identified and programs initiated to fill the needs, planners realized that the military lacked sufficient numbers of weather forecasters to support operations in widely varied environments all over the world. Deserts, rain forests, arctic zones, vast continental regions, mountains, the Pacific Ocean and its islands, the Atlantic Ocean, the Mediterranean Sea—American soldiers, aviators, sailors, and marines were sure to fight in all these places and would need experienced meteorologists to guide commanders in their decisions.

Weather has concerned US military leaders since the first days of our armed services, but units in the field or at sea were on their own to assess weather conditions as best they could. Even during World War I they had few resources other than the commander's experience to make decisions about the weather.

During the peacetime interlude following World War I, commercial airline traffic and air mail service grew rapidly, but as weather-related accidents mounted, weather forecasting became increasingly important.[1] The US Post Office Department inaugurated US Air Mail Service on the Washington-Philadelphia-New York route May 15, 1918, with army pilots and aircraft.[2] These early air mail pilots flew by the seat of their pants. The first official aviation weather forecast was made by the US Department of Agriculture's Weather Bureau on December 1, 1918.[3]

The army flew the mail for four months, until August 1918, and then the service was taken over by the Post Office Department's Aerial Mail Service, which flew the mail in government-operated aircraft. This arrangement lasted until 1927, when Congress directed the US Post Office Department to contract with commercial air carriers to develop air mail routes.

Then in 1934, to resolve a dispute with the commercial airlines, President Roosevelt announced that the Army Air Corps would begin flying all of the nation's air mail for a twelve-week trial period.[4] This program was a disaster, with twelve army pilots killed in sixty-six crashes during the first four months of the program. The Air Corps blamed the disasters on lack of navigational support from the Signal Corps, but many of the accidents were the result of inexperienced pilots and lack of training.

A presidential board appointed to investigate the crashes recommended that the Signal Corps provide three more meteorological companies to support the Air Corps. Because of the army's outdated organizational structure, most weather personnel were assigned to the Signal Corps even though the Air Corps had the greatest need for weathermen. When the Air Corps requested additional forecasters from the Signal Corps in 1935, Major General James B. Allison, the chief signal officer, refused to authorize any additional weather billets.[5] He considered weather forecasting to be a minor component of his responsibilities that was overshadowing the Signal Corps' main mission of providing communications support to army units in the field. In 1937, after sixteen months of wrangling, the army assistant chief of staff directed the Air Corps to provide weather services to itself and to ground forces at division level and higher. The Army Air Corps finally had the authority to provide its sorely needed weather services. Signal Corps meteorological personnel were given the opportunity to transfer to the Air Corps, and many did. By late 1939, the number of weather officers and enlisted men in the Air Corps had doubled to 30 officers and 388 enlisted men.

In the spring of 1940, as Hitler's armies advanced across Europe, President Roosevelt authorized a massive buildup in America's military strength even though the United States had not yet entered the war. Captain Arthur F. Merewether, head of the Training and Operations Division of the army's Washington, DC, Weather Central, realized that the Army Air Corps was still woefully short of qualified weather personnel.[6] Until then, Air Corps weather trainees had been West Point graduates or pilots before being sent to either Caltech or MIT for weather training. Captain Merewether arranged for a new aviation cadet training program to be taught at the nation's five meteorological schools. To handle the expansion, programs at UCLA and the University of Chicago were accelerated, and full-scale military meteorological programs started at these schools in the fall of 1941. Army recruiters vigorously sought qualified candidates for the rigorous nine-month graduate-level programs.

~

IN EARLY 1942, Bob Reid, a third-year engineering student at the University of Southern California became concerned about his draft status.[7] His college deferment would last through the next school year and give him time to finish his senior year at USC, but it also specified that as soon as he graduated he would be drafted. He knew he would have little or no choice of assignment.

He visited an army recruiter to see what options were available to him as a young man with over two years of college credit. The recruiter told him that the Air Corps urgently needed meteorology officers. If he volunteered now, he could complete his junior year at USC and would be called to active duty in June 1942. The army guaranteed him a commission as a second lieutenant in the Army Air Corps and that he would be enrolled in the new meteorology program.

Bob discussed the options with his family. Should he wait a year and finish his degree with the near certainty of being shipped to the front lines as cannon fodder or volunteer now and put his talents to work as a weather forecaster? The answer was obvious to a very bright young man with a gift for mathematics and technical subjects.

Bob Reid volunteered and was inducted into the army June 15, 1942. After three months of basic training, he began meteorology courses at UCLA in September. He was one of about two hundred cadets in Class 5 of the Army Air Corps Weather Officer Training Group, where he studied physics, mathematics, and meteorology under some of the best meteorologists of the day—Jacob Bjerknes, Jorgen Holmboe, Morris Neiberger and Joseph Kaplan.[8] During training, Bob Reid became good friends with Don Pritchard, a fellow cadet in Class 5. Reid described Pritchard as "much more outgoing and athletic than I, having been a member of the Cal Tech football team."[9] The quiet, introspective Reid recognized Don Pritchard as "a very intelligent and congenial person," traits that also described Reid. They probably formed the main brain trust of Class 5. Little did either know that their lives would be intertwined during the war and in their careers afterward. They finished their course in August and received commissions as second lieutenants in the Army Air Forces September 6, 1943.

Weather forecasting at that time was based on what was known as the Norwegian cyclone model, a technique developed over the prior quarter century at the Geophysical Institute, University of Bergen, Norway.[10] It was founded on research by Wilhelm Bjerknes and his students, including his son Jacob—who then was on the UCLA faculty teaching US Army officers, including Lieutenant Bob Reid.

According to the Norwegian theory of cyclone development, two air masses, one warm and one cold, come into contact with each other along what is called a frontal boundary. In the initial stage of the cyclone development, a wave of low pressure forms along this frontal boundary when the warm air mass overtakes the cooler air mass. The warm air rides over the cool air, creating a sloping boundary and moves poleward (northward in the northern hemisphere). A disturbance develops in the upper level of the atmosphere and moves toward the surface low. Precipitation begins ahead of the poleward advancing front. The surface low-pressure area deepens and becomes better defined and eventually merges with the upper-level disturbance. When the cold front reaches the western portion of the warm front, the warm air is separated from the cyclone's center at the earth's surface (occluded front) and the cyclone begins to dissipate. The Norwegian cyclone model is still the basis of today's weather forecasting and remains remarkably unchanged.[11]

During the spring and summer of 1942, army planners realized that the amphibious landings that were already in the planning stages would require specialized forecasts of the coastal ocean environment. The meteorology course at UCLA was producing forecasters to predict large-scale weather development, but the army also needed a smaller number of weather officers to make localized forecasts of the ocean environment in the areas where soldiers and marines would go ashore during amphibious landings.

While Reid and Pritchard were completing their basic weather course, the army started an advanced course in coastal weather and ocean forecasting. This course, taught at the Scripps Institution of Oceanography and UCLA, was designed to produce ocean forecasters with the skills necessary to predict the coastal environment: ocean wave heights and periods, surf, tides, and marine fog for amphibious landings. The top eight or ten students in the UCLA meteo-

FIGURE 1

Walter Munk (*left*) and Harald Sverdrup in the George H. Scripps Memorial Marine Biological Laboratory building at Scripps Institution of Oceanography, circa 1940. Courtesy Special Collections & Archives, University of California, San Diego.

rology program were selected to attend the Scripps Coastal Weather and Ocean Forecasting course.[12] Cadets from other services, including the US Navy and Marine Corps also were selected for these courses.

Again, the army provided the very best instruction available. Dr. Harald Sverdrup, director of Scripps, taught courses on waves and tides. His protégé, Walter Munk, also taught courses on waves and other physical oceanographic subjects (see Figure 1).

Sverdrup has often been referred to as the "father of modern physical oceanography." Before his time, the subject was rarely taught, but when it was, it was taught as a hodgepodge of mathematics and physics, not as a unified discipline. Sverdrup's crowning achievement, the classic text *The Oceans: Their Physics, Chemistry and General Biology,* published in 1942, established Scripps as an international oceanographic research institution and is still used today.[13] The military considered Sverdrup's book so valuable to the war effort that it was banned from distribution abroad until the war ended.

Before World War II, conditions at sea were described using a qualitative, nonmathematical method that expressed wind speeds on a scale of one to twelve in what was known as the Beaufort Scale. The scale was originally devised by British Rear Admiral Sir Francis Beaufort in 1805 based on his observations at sea. Waves were described in narrative terms. For instance, a gale was Beaufort Force 8, which meant it had wind speeds of thirty-four to forty knots. A gale had "moderately high waves of greater length [greater than Force 7]; edges of crests begin to break into spindrift. The foam is blown in well-marked streaks along the direction of the wind."[14]

The amphibious operations being planned would require accurate, site-specific forecasts of ocean conditions (wave heights and periods, direction of propagation, current speed and direction) at the location of the landing, as well as the evolution of those conditions for several days before the invasion. Although the qualitative Beaufort Scale is reasonably descriptive of the appearance of storm waves at sea, it is totally unsuited for any kind of quantitative calculations. The Beaufort Scale applies to wind and waves on the open ocean, but military planners also needed to know conditions in the nearshore zone. They needed a mathematical method they could use to calculate wave heights based on wind speeds and then propagate the deepwater waves into shallow water at a proposed amphibious landing site.

~~~

IN AUTUMN 1942, Operation Torch was being planned to invade North Africa by landing troops near Casablanca, Morocco, where wintertime sea conditions are notoriously rough. Captain Harry R. Seiwell, an oceanographer at the Army Air Forces Research Center in Washington recognized the army's need to forecast sea and surf conditions not only for Operation Torch but for other amphibious landings that were sure to follow.[15] Before the war started, Dr. Seiwell had been an oceanographer on the faculty of the Woods Hole Oceanographic Institution in Massachusetts, where his research centered on identifying and tracking water masses. He knew that ocean forecasting was outside his areas of
~~~

expertise, so he obtained funding to bring Walter Munk from Scripps to Washington in September 1942 to try to devise a wave and surf forecasting method.

Munk spent the month assembling all the data he could find to develop empirical relationships to relate sea, swell, and surf to wind conditions.[16] At the end of September, not completely satisfied with his results, he told Captain Seiwell that he thought it was possible to develop a workable method, but he needed Dr. Sverdrup to help him make sense of the scattered data. Seiwell asked Sverdrup to come to Washington to work on the problem. Munk and Sverdrup pored over the data, and by the end of October Dr. Sverdrup was satisfied that the method could be used to predict waves for Operation Torch.

Dr. Munk said they were not sure whether their "prediction rules" arrived in time to be used for the invasion or not. The ocean was unusually calm for that location at that time of year, however, and the American forces landed successfully at Casablanca in North Africa November 8, 1942.

After completing their work for Operation Torch, Sverdrup and Munk returned to Scripps and during the first half of 1943 worked together to refine their wave forecasting method. They knew they had developed it hurriedly to meet an immediate wartime need and that it could be improved. Seafarers and scientists alike had long recognized that winds create waves, but some of the theories of wave creation seemed problematic to these two distinguished scientists and did not lend themselves to mathematical calculation of wave properties.

Sverdrup and Munk theorized that waves were generated by the frictional stress between the wind and the sea surface as winds blew over the ocean. Think of someone sliding a hand lightly over a loose tablecloth. The friction between hand and cloth causes the cloth to wrinkle in front of and under the hand. The tablecloth doesn't move very far, but wrinkles (waves) are created and then disappear soon after the hand (wind) moves on. The physics in these two cases are different, but the effects are similar. Figure 2 illustrates common wave nomenclature.

Wave height is the vertical distance (trough to crest) through which the water surface rises and falls as a wave passes. Imagine a person standing in the ocean at the beach. As a wave crest approaches, water rises to the armpits, then drops to the knees as the wave trough reaches the person. In this example, the wave height is the distance between the person's knees and armpits. Wave length is the horizontal distance between two crests (or troughs) or any other similar

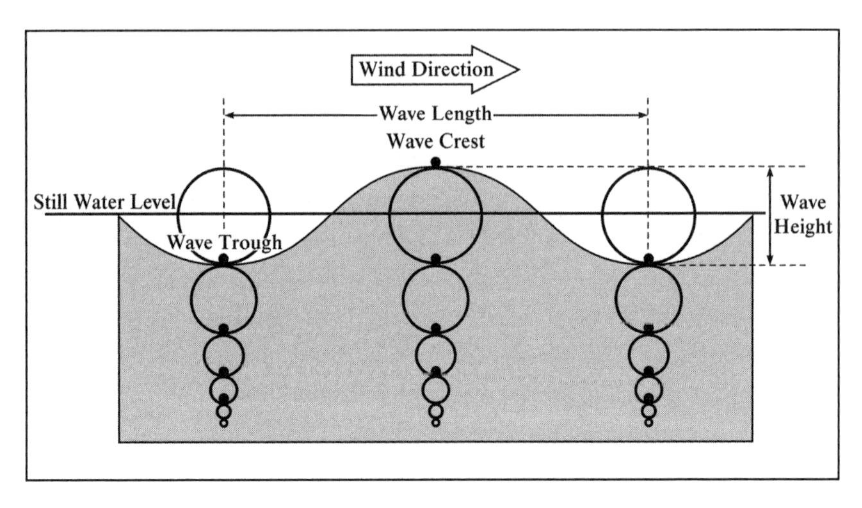

FIGURE 2

Sea-wave nomenclature. In deep water, water particles, or debris in the water, follow circular paths as a wave passes. In one wave passage, a particle traces a complete circle, then returns to its original position. Note that the circles become smaller with depth. Wave motion disappears deeper in the water column. In shallow water, where the wave paths reach bottom, the orbits become elliptical in shape.

points on consecutive wave profiles. The wave period is the time (in seconds) between two successive waves, or the time between when the person's armpits first get wet until they get wet the next time.

As a wave passes a point in deep water, water particles, debris suspended in the water, or a fishing float on the surface follow circular paths (in the vertical plane). The circles become smaller as one goes deeper into the water and at a depth equal to approximately one half the wave length become so small that they are nearly undetectable. This is why submerged submarines can pass under the most severe storms at sea (hurricanes, for instance) without the submariners even knowing there is a storm overhead.

Sverdrup and Munk developed sets of mathematical equations based on their analyses of the available data sets. They related wave heights to the wind speed over the sea's surface, wind fetch (distance over which the wind has blown), and wind duration (time the wind blows at that speed and direction). With a plausible and quantifiable explanation of how waves are generated and how they grow higher if the wind continues to blow, they tackled the problem of what happens to a wave once it is formed.

Waves travel away from the windy area where they are created and can travel hundreds, even thousands of miles to crash on distant shores. When waves propagate out of the area where they were generated, they are known as swell and can appear suddenly on a far-off beach on a windless day. Sverdrup and Munk added more mathematical equations to their wave generation model to quantify their thoughts on wave propagation. They now turned their attention to designing simple charts, graphs, tables, and procedures that a forecaster could use to predict wave and swell heights caused by a nearby storm, or by one half an ocean away.

To use this method, a forecaster entered the pertinent information from weather maps (storm size, wind speed and direction, distance and direction from the storm to the beach of interest) on the worksheets and graphs. Then he calculated wave heights generated by the storm and forecasted when and where the swell would reach shore. The method also worked to forecast locally generated waves for storms in the immediate vicinity of the beach for which he was forecasting.

Wave heights calculated by this method were for waves in deep water and did not take into consideration changes in wave characteristics when the waves reached shallow water. As a wave travels into shallower water, the wave orbits eventually reach the bottom, and in oceanographic jargon, the wave "feels" bottom. As the wave continues shoreward, the orbits become elliptical, and the wave slows down.

To visualize these shallow-water transformations, imagine two successive waves approaching a beach. The first one feels bottom and slows down, while the second continues shoreward at its original speed. The second wave begins to catch the first one, shortening the distance (wave length) between them. Eventually, the wave breaks in the surf zone when it reaches water that is 1.25–1.5 times as deep as the wave height. This is an oceanographic rule of thumb that wave forecasters use to estimate the depth along a beach at which waves will break.

In the spring of 1943, the wave forecast method was complete and ready for more rigorous testing. Would it work in operational situations, and if so, how well? To answer these questions, Major Harry Seiwell (having recently been promoted from the rank of captain) arranged for Dr. Sverdrup to select eight newly minted weather officers from Class 4 of the UCLA meteorological training course to report to Scripps to evaluate the method.

These men became the first class in the new Coastal Weather and Ocean Forecasting course at Scripps. Two of these officers, Lieutenant Charles C. Bates and Lieutenant John C. Crowell, were destined to work together less than a year later to produce the all-important beach forecast for D-Day.

Dr. Sverdrup wanted to see if the eight student oceanographic meteorologists could start with actual weather maps of an oceanic area and produce forecasts of waves at a selected location. Accurate wave measurements had been made in the 1930s at a beach in Casablanca during construction of a breakwater. Without the forecasters knowing anything more than what they could glean from weather maps of the Atlantic Ocean at the time the wave measurements were made, the students filled in the pertinent data in the Sverdrup-Munk worksheets and did the calculations to "hindcast" waves at the Casablanca beach. A hindcast uses the same methods as a forecast, but historical weather data are used instead of the projected weather conditions used in a forecast. The results were excellent. This confirmed the accuracy of the method and its effectiveness when used by relatively inexperienced forecasters. An analysis conducted after D-Day showed that when wind forecasts were accurate, this method predicted wave heights to within one foot of actual wave heights 83% of the time.[17]

The Sverdrup-Munk wave forecasting method was quickly added to the curriculum for succeeding Coastal Weather and Ocean Forecasting courses to train more oceanographic meteorologists for the amphibious landings that were already being planned. A declassified version of the method was published after the war.[18] Walter Munk admits that their method was crude, even by 1946 standards, and is totally inadequate today, but it was developed quickly to fill an urgent wartime need for surf forecasts, and it worked.[19]

Dr. Blair Kinsman, in his comprehensive textbook on waves, *Wind Waves: Their Generation and Propagation on the Ocean Surface,* wrote that despite the shortcomings of the Sverdrup-Munk forecasting method, "there are some thousands of World War II veterans alive today who would have been dead in the surf had Sverdrup and Munk not done their best with what they had."[20]

Once America entered the war, most activities, civilian or military, that were in any way associated with the war effort were subject to strict security restrictions. The work of the two foreign-born scientists, Sverdrup (Norwegian) and

Munk (Austrian), required high-level government security clearances.[21] Both had many relatives living in Europe during the war, and some Scripps employees and local La Jolla residents were never quite convinced of their loyalty to the United States. Whether due to real concerns or professional jealousies, they reported "suspicious activities" to the FBI. Both Sverdrup and Munk were granted clearances, only to have them revoked, then reinstated. Sverdrup, especially, had repeated problems with his clearance. Remarkably undeterred, they continued their work through the war years under army auspices and later were cited by naval officers for saving countless lives during the war.

~

GRADUATES OF THE basic weather forecasting program at UCLA and the Coastal Weather and Ocean Forecasting Program at Scripps were sent to all theaters of the war. In preparation for the invasion of Europe, the army established the US Army Assault Training Center in the coastal town of Woolacombe, Devon, on the southwest coast of England. A weather station, designated Weather Station WZ, was created at the center. Weather forecasting for the training exercises was nearly as important as for the actual invasion, so a succession of new UCLA/Scripps meteorologists and oceanographers was sent to Weather Station WZ to forecast and train in their specialties.

Woolacombe, located on Barnstaple Bay and facing the Celtic Sea, was selected for the training center because its beach resembled the Normandy beaches in many ways. Even the hinterland was similar to that at Normandy. Tide range at Woolacombe is about twenty feet, as at Normandy, and the beach slope is gentle, similar to the gradient at Normandy's Omaha Beach. The beach is about three miles long, which provided sufficient space to conduct several landing exercises simultaneously. It faces northwest, giving it a similar exposure as the northward-facing Normandy beaches. The distance from Woolacombe across the Celtic Sea to the coast of Ireland is roughly the same distance as from Normandy across the English Channel to the coast of England. This gives Woolacombe about the same wind fetch and sheltering from storms as the Normandy beaches. The main difference is exposure to the open Atlantic Ocean. The Cherbourg Peninsula protects Normandy from waves approaching from the west, but Woolacombe lacks this protection.

FIGURE 3

A DUKW, an amphibious two-and-a-half-ton truck, brings supplies ashore on a Normandy Beach, June 11, 1944. Courtesy Naval History and Heritage Command, "80-G-252737 Normandy Invasion, June 1944." National Archives ID 252737.

The beachfront land was expropriated from the local residents, and they were moved elsewhere. To increase the realism of the training, the army constructed obstacles offshore similar to those that the Germans had erected at Normandy. Training began at the Woolacombe Center September 1, 1943, and by D-Day, tens of thousands of soldiers had trained there. They learned the techniques of landing on a heavily defended enemy beach, and their leaders developed tactics for the invasion.

Lieutenant John Crowell was assigned to the center from October through December 1943.[22] The weather station commander made good use of his training and put him to work improving wave observation techniques. It was important to consistently estimate wave and breaker heights with an accuracy of one foot or less, a difficult task when standing on shore looking seaward. They decided that the most reliable way to measure wave heights and periods was to go offshore each day in a navy landing craft and make the measurements while being tossed about by the very waves they were measuring.

Lieutenant Crowell, with one of his enlisted men at Weather Station WZ and a driver spent some seasick days in a DUKW (amphibious truck, see Figure 3), measuring wave heights and trying to determine how the ungainly vehicle handled in the Woolacombe Beach breakers. Their wave measurement equipment consisted of a long pole and line, much like a fishing pole, with different colored ribbons tied at one-foot intervals along the line.[23] As the DUKW driver attempted to hold the craft in position seaward of the breakers, one of the weather observers stuck the pole over the side, lowered the line until a weight at the end of the string touched bottom, then observed the wave heights as the waves passed the string. He called out his observations for recording to the other observer, who usually was retching in the bottom of the boat. It was a crude method, but it worked reasonably well and the data were useful for verifying wave forecasts, for studying wave shoaling (how wave heights change when they reach shallow water), and for estimating the limits of the DUKW's operability. "Most of all," Lieutenant Crowell said, they gained "a first-hand appreciation of what soldiers would face in a landing craft as they stormed a beach."[24]

~~~

AFTER COMPLETING THE UCLA meteorological course, Lieutenant Bob Reid, Lieutenant Don Pritchard, and eight other weather officers from their class were selected for Class 2 of the Coastal Weather and Ocean Forecasting course.[25] When they finished, Reid and Pritchard parted ways, not knowing that they would meet again on the beach at Normandy. Reid was stationed at air bases in California and North Carolina to forecast weather for training operations, then was reassigned to the 21st Weather Squadron of the 9th Air Force in England.

On the morning of January 30, 1944, Reid boarded the RMS *Mauretania,* bound for England. It was part of a convoy of troop ships escorted by destroyers that zigzagged their way across the Atlantic to elude the German U-boats lurking beneath the gray wintertime sea. Mountainous waves battered the convoy during much of its two-week crossing of the North Atlantic and kept the
~~~

soldiers perpetually seasick. But they were glad to trade their meals to Neptune in exchange for high seas that hindered the German submarines' ability to hunt surface ships. They arrived safely in Liverpool February 14, 1944.

Colonel Thomas S. Moorman Jr., commanding officer of the 21st Weather Squadron, briefed the newcomers on their assignments, and Bob Reid learned that his friend Don Pritchard was already stationed in London. After a week of the army's usual hurry-up-and-wait routine, the newly arrived weather officers and enlisted men were assigned to various air bases in London, other parts of England, and Wales.

Lieutenant Reid and another second lieutenant were sent to the isolated Weather Station WZ at Woolacombe. Their quarters were on the second floor of a drafty rooming house, and they were cold most of the time. Coal was rationed for civilian use and the proprietress only had enough to heat the lower floor. They ingratiated themselves to her (and kept warm) by bringing coal from the weather station to heat the old house.

When Reid arrived at the center in February, he was assigned to forecast beach conditions for training exercises that later proved to be dress rehearsals for D-Day. Colonel Moorman asked Reid to observe and record wave heights and periods at Woolacombe to evaluate the accuracy of the Sverdrup-Munk wave forecasting procedure.

It was common knowledge that Operation Overlord, the invasion of mainland Europe, was being planned, but no one knew when or where it would happen. Even so, Reid sensed that he and Pritchard might play significant roles in the forthcoming invasion.

3

LOW SPRING TIDE AT NORMANDY

AFTER THE UNITED STATES ENTERED World War II, the leaders of the Allied nations held a series of conferences to plan the liberation of western Europe from the Germans. At the first meeting, in January 1943 in Casablanca, President Franklin Roosevelt, British prime minister Winston Churchill, and other American and British officials and military officers decided it was time to begin detailed planning for the invasion of Europe. They established the position of chief of staff to the supreme allied commander (COSSAC), even though the Supreme Commander had not been appointed yet.[1] British Lieutenant General F. E. Morgan was appointed chief of staff, with Brigadier General R. W. Barker, an American as his deputy. General Morgan's mission was to assemble a staff of military experts and start planning the myriad details of this grand assault on the European continent.

Planning began immediately and continued during the remainder of 1943. Among the innumerable commands and organizations COSSAC formed and staffed, in November 1943 it established a meteorological group to assist the supreme commander in all matters pertaining to weather. It was staffed initially with only two men: a British civilian, Dr. James M. Stagg, was the chief, and an American Army Air Forces officer, Colonel Cordes F. Tieman, was his deputy. They had no clerical or secretarial help of any kind. When General Dwight D. Eisenhower was appointed supreme commander in January 1944, his meteorological advisors had already been on the job for two months. A few weeks after General Eisenhower took command, COSSAC was renamed the Supreme Headquarters of the Allied Expeditionary Force (SHAEF).

By the time Dr. Stagg was assigned as chief of the meteorology group, the COSSAC staff had decided that the invasion had to occur at the lowest tide of the month so skippers of the landing craft could avoid obstacles that the Germans had placed in the surf zone. Aerial photographs had revealed several different types of obstacles that were visible only at low tide. One type, called a hedgehog or tetrahedron, resembled a gigantic children's jack about five feet high, made of railroad rails welded together or logs lashed together. Another, called a Belgian gate, did in fact look like a gate more than seven feet high. Steel rails held the gates in place and ran toward shore, anchoring them very firmly in place. No landing craft could withstand a collision with one of these gates. Lining miles of the French beaches were log ramps as well as logs simply driven into the ocean bottom at intervals of a few feet. Many of these obstacles were mined, but even if they were not, they could impale and sink any landing craft that hit them. The low tide needed to be in early morning to give engineers thirty minutes after daylight to clear lanes through the obstacles before the first wave of landing craft left for the beach. The first wave would be followed by successive waves of landing craft as the day wore on.

The lowest tides occur twice a month, when the moon is full and again on the new moon. The air force planners wanted a full moon so bomber pilots could see their targets well enough to bomb them before the landings started. The navy also wanted a full moon so their spotters could see where the shells from their big guns were hitting. The Army Air Forces needed full moonlight for several hours before daylight so glider pilots could find clear spots to land behind German lines. The unpowered gliders only got one shot at a landing; they could not abort and go around for a second attempt. Moreover, to initiate the invasion, paratroopers from three airborne divisions (two American, one British) were to parachute in behind the German forces shortly after midnight. They also needed moonlight as they drifted earthward to pick a safe landing spot among the trees, houses, and bocages in the French countryside. These weather restrictions limited the available invasion dates to a two- or three-day period early in the month or near mid-month in each of the months of May, June, and July.

Bocages are pastures or cultivated fields surrounded by hedgerows of raised earth with trees, hedges, and sometimes stone fences atop them. Even tanks found cross-country travel impossible in hedgerow country and had to improvise ploughs and bulldozer blades to cut through them.

Dr. Stagg asked members of General Eisenhower's planning staff for a list of the meteorological parameters they considered important for D-Day. Each wanted "quiet weather," but what did that mean? He had to rephrase his question to, "What are the least favorable conditions in which your forces can operate?"[2]

At first the army's requirements seemed the easiest, until it added no fog or heavy mist in the glider and airborne landing zones, less than 6/10 cloud cover, no clouds below three thousand feet, and winds less than twenty miles per hour in the drop zone.

The air forces had different criteria for fighter and bomber aircraft and even for bombers on different missions. The high-level bombers wanted no clouds below ten thousand feet, but the low-level fliers wanted clouds down to their flight altitudes to provide security from Luftwaffe fighter planes but no clouds from there to the ground.

The navy's requirements, however, outweighed most of the others. Everything would come to naught if winds were too high or seas too rough for their landing craft to get the soldiers to the beaches. Even no wind could be a problem because heavy fog would be likely. So the navy wanted winds less than ten to twelve miles per hour and visibility of at least three miles so their guns could be sighted on shore targets and spotter aircraft could see where the shells hit. It was "highly desirable" that there be no prolonged periods of high winds over the English Channel or its approaches from the Atlantic for several days prior to D-Day so that any swell or storm waves had time to dissipate.

The combined air forces (US Army Air Forces and British Royal Air Force) also stipulated necessary weather conditions in southern and eastern England (in addition to the English Channel and French coast) because their aircraft had to return to bases there to refuel and rearm after flying missions to Normandy.

With this overwhelming (and sometimes conflicting) list of desired conditions in hand, Stagg went back to the planners and asked them to trim their criteria to a manageable list. All were reluctant to reduce their requirements or be made second priority. He finally decided to make his own list, "which, remarkably, I was never asked to produce," he said. It should be remembered though, that Stagg did not have to make the decision about the operability of the weather.[3] That responsibility rested solely on General Eisenhower. Dr. Stagg only had to forecast the weather as accurately as he could. He estimated from climatological data that the odds of satisfying all the services' wants were about fifty to sixty to one if D-Day had to occur on a full moon, which it did.

~~~

ABOUT THE TIME THAT Dr. Stagg was appointed to lead the meteorological group, the Allied High Command made the British Royal Navy responsible for all military matters pertaining to the sea. The Royal Navy was shorthanded and the US Navy had no spare aerologists (meteorologists) for this duty, so Colonel Moorman, commander of the US Army Air Forces 21st Weather Squadron, offered the assistance of his uniquely qualified oceanographic meteorologist, Lieutenant John Crowell, to the director of the Royal Navy Meteorological Service, Captain Lloyd Garbett.[4] Crowell was nearby at Woolacombe, so during November and December 1943 he made several trips to London to confer with British and American senior officers on sea, swell, and surf forecasting.

Of particular concern was the accuracy of the Americans' Sverdrup-Munk wave forecast method versus the Suthons method, a more empirical technique based on observations the British devised hastily for Operation Torch. Crowell wrote, "How could we find out if forecasts using the two methods were similar or different? Should the wave forecasters for the invasion use one of them or both? Was there a way to test the two methods and calibrate them?"[5]

To answer these questions, Colonel Moorman and Captain Garbett sent John Crowell and his British counterpart, Commander Dick Burgess, to North Africa. They were to interview some of the planners of landings at Morocco and Sicily to (1) learn how well wave forecasts correlated with actual conditions during landings of Operation Torch in Morocco in November 1942; (2) determine if there was useful information to be gleaned about wave forecasting techniques used for invasions in Sicily in July 1943; and (3) arrange for wave observations in North Africa and the Azores to compare and test the Sverdrup-Munk and Suthons wave forecast methodologies.[6]

Crowell and Burgess left England in the middle of the night on December 27, 1943, and flew to Gibraltar. Burgess continued to Casablanca, while Crowell went on to Algiers. Officers in the US Mediterranean Headquarters in Algiers told Crowell that accurate forecasts of local winds and waves were of paramount importance and needed to be made at least two days before the planned oper-
~~~

ation. They were not concerned with swell from more distant storms. Experience with landing craft at Salerno, Italy, confirmed Crowell's own experience at Woolacombe with these awkward craft, whose safe operation seemed to be limited to waves no higher than three or four feet.

On December 31, 1943, Crowell flew to Casablanca and arrived just in time to meet Burgess for a New Year's Eve party. During the next twelve days, they arranged to have observations made of waves and swell that reached the Moroccan coast from storms in the North Atlantic.

To obtain these data, Commander Burgess got permission to paint stripes at one-foot intervals on the hull of the Vichy French battleship *Jean Bart,* which had been damaged by American dive-bombers during Operation Torch and now lay grounded in the harbor. They arranged for two groups of observers to record the heights of waves as they passed the ship, the end of a long breakwater, and at an offshore buoy.

Crowell then flew to the Azores and persuaded the assistant director of a Portuguese oceanographic station to forward reports of wave observations to London. He wanted to compare wave heights and periods and the arrival times of swell at the Azores with those from the same storm at Casablanca as a test of their forecasting methods. Wind speeds in the storms, used as input to the wave forecasting procedures, would come from weather maps of the North Atlantic.

The "quick and uncertain" comparison that he and Commander Burgess made showed that forecasts of swell heights and periods were similar using either the Sverdrup-Munk method or the Suthons method.[7] They settled on the Sverdrup-Munk method because it was quicker to use.

A new group, known as the Swell Forecast Section, was formally organized February 1, 1944, within the Royal Navy Meteorological Centre to forecast waves for Operation Overlord. Commander John Fleming of the Royal Navy was appointed to lead the section. Since Commander Fleming could not get the meteorological help he needed from either the Royal Navy or the US Navy, Crowell, who had been assisting him for several months on an as-needed, temporary assignment was officially transferred from Woolacombe to London to put his skills to work forecasting swell and surf for D-Day.

~~~
~~~

AFTER BATES COMPLETED the Coastal Weather and Ocean Forecasting Course at Scripps, he was ordered to duty with the 8th Weather Squadron stationed at the Stephenville Air Force Base in northern Newfoundland to forecast weather for military passenger flights between the United States and the United Kingdom. He remained in contact with his friend John Crowell and was aware that Crowell was involved with preparations for the all-important D-Day forecast. One night when Bates was on duty in the weather office, Colonel Moorman, who was aboard a flight that stopped at Stephenville to refuel, came by the office to inspect the weather maps prior to the crossing to Scotland. Bates was anxious to play a more active role in the war effort and seized the opportunity to lobby the colonel for a transfer to London. In short order, on February 10, 1944, he was assigned to ninety days' temporary duty with Moorman's command and soon joined Lieutenant Crowell in the Swell Forecast Section.

Charles Bates, John Crowell, and a British meteorologist, Instructor Lieutenant H. W. Cauthery of the Royal Navy made up the Swell Forecast Section at the Admiralty. They were assisted by two US enlisted men and two members of the Women's Royal Naval Service.

The five men and two women worked in a dingy, unheated room some ten feet by sixteen feet in size. Their unmarked work space, in the subbasement of the new Admiralty Building, adjacent to the Horse Guards Parade in London, was dark and cramped.[8] The Admiralty Meteorological Service, the Royal Navy's D-Day weather forecasting unit with whom they worked closely, occupied two rooms in the very bowels of the Citadel, a bombproof facility about two hundred feet from the Admiralty Building. It was entered by a long ramp that descended two floors underground and had bank-vault doors that could be closed quickly in case of a poison gas attack. Thus the men and women of the Swell Forecast Section were only a short walk from the Admiralty's Forecast Section and had access to its excellent weather data and maps.

Charles Bates wrote, "The objects of the [Swell Forecast] Section were to develop the technique of forecasting sea, swell, and surf, and to provide forecasts on the basis of this technique for the invasion of Europe. The technical problems facing the section were four in number: (1) to forecast the height and period of ocean swell coming from the Atlantic; (2) to determine the extent to which this swell would penetrate the English Channel; (3) to forecast the height and period of waves caused by local winds in the channel; and (4) to study the effects of shallow water, tidal currents, and coastal irregularities on waves."[9]

This could be a full university research program that would employ a number of investigators and graduate students for several years. The section had three young scientists and three months to do it. Not making matters any easier, security restrictions prevented them from conferring with any of their old professors or colleagues in the United States. They were on their own, and they had to be correct.

The information on waves that the section's officers needed to verify forecasts and develop surf predictions did not exist, so they had to get it themselves. They set up a wave observation network that extended around the southern coast of England (from west to east): from Woolacombe, southwest around Land's End, along the southern English coast, and past the white cliffs of Dover to East Anglia (Map 2). They arranged for lookouts at fifty-one Coast Guard stations to estimate wave heights and periods three times a day. These observations were encoded and sent to the British Admiralty and eventually reached the Swell Forecast Section.

Their plan was to forecast waves and swell for these British beaches and then use the observations to evaluate the accuracy of the forecasts. Since the exposures of the British and Normandy beaches were essentially mirror images of one another, they reasoned that if they could establish the proper relationships for winds blowing onto the English coast, similar relationships would exist when the winds were reversed and blew onto the French coast.[10]

Another rudimentary but effective way to evaluate the accuracy of the forecasts was through aerial photographs of the French coast. From the photographs they could determine where the waves began to break, and from tide data and charts of beach profiles they could estimate the water depth where the waves broke. Then, using the oceanographers' rule of thumb that waves break when they reach water that is 1.25–1.5 times the wave height, they calculated the heights of the waves.

The first technical problem they tackled was to decide if detailed swell forecasts would be required for the invasion beaches. After studying swell observations at Land's End for several weeks they concluded: (1) that the Sverdrup-Munk method accurately predicted swell in the open portions of the channel

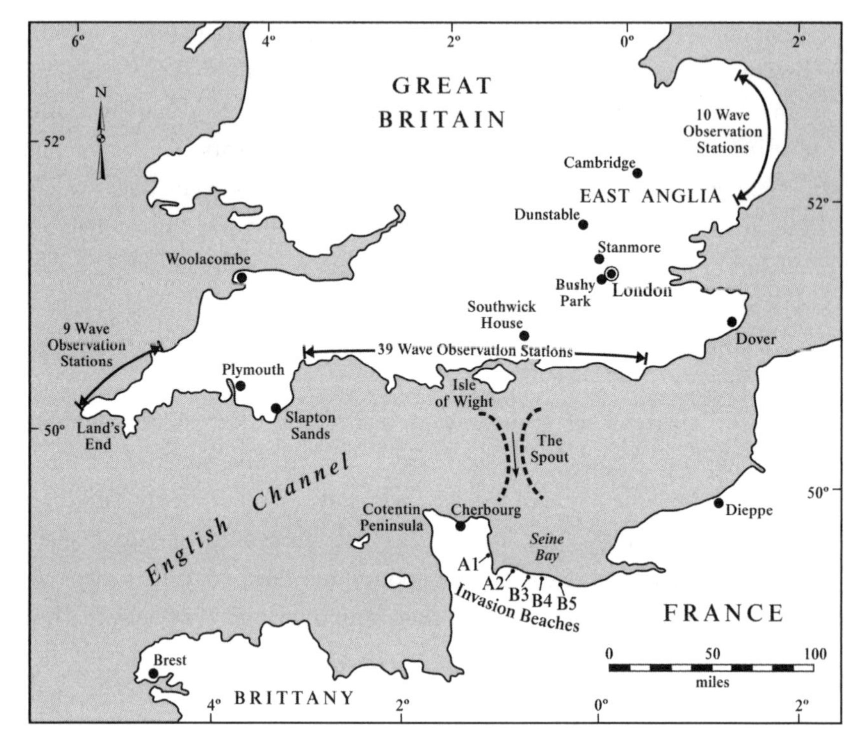

MAP 2

Wave observation stations set up by the Swell Forecast Section. Fifty-one of these stations provided useful data. Adapted from Bates, "Sea, Swell, and Surf Forecasting for D-Day and Beyond," 8.

and (2) that during summertime weather conditions, it would be very unusual for significantly high swell to reach the invasion beaches. Therefore, they reasoned, swell would not be a problem at the Normandy beaches and they only had to concern themselves with forecasting waves and surf from local storms within the channel.[11] Their first two problems were solved.

Waves change as they travel from deep water into shallow water. In water that is deeper than twice the wave length, which is most of the ocean, even very large (and long) waves such as in a hurricane do not reach bottom. When the wave reaches shallow water, it "feels bottom" and becomes higher, steeper, and shorter, a process known as shoaling.

When waves approach a beach at an angle, the end of the wave that feels bottom first begins to slow down, while the rest of the wave continues shoreward at its original speed. As points along the wave front feel bottom, the front slows down and bends, so that the wave strikes the beach parallel to shore. This is called refraction. When a wave hits an obstruction such as a headland or breakwater, the portion that strikes the obstruction is blocked, but the remainder of the wave continues on its original path. After the wave passes the obstruction, it begins to spread laterally behind the obstruction, a process known as diffraction. Tidal currents that oppose or run with waves, or at angles to them, alter wave heights. Also, large rocks offshore and coastal irregularities change the deepwater waves before they reach a beach. Waves at the observation stations along the southern English coast and at Normandy were subject to these effects.

Lieutenants Bates and Crowell had studied these processes at Scripps and knew they had to be considered for Normandy, but a comprehensive theory for doing so had not been published. Again they were on their own. They decided to use the empirical approach: find mathematical relationships between the observations of actual wave heights and the forecasted waves in deep water. The effects of shoaling, refraction, diffraction, tidal streams, and several smaller influences would be automatically incorporated into any relationships so derived. Bates wrote, "If the necessary relationships between computed and observed data could be established for winds blowing onto the English coast, it was hoped that similar relationships would exist for winds blowing onto the French coast."[12] For the actual D-Day forecasts, they would compute deepwater wave heights in the channel using the Sverdrup-Munk method, then use their empirical relationships to compensate for shallow-water effects and predict surf at the Normandy beaches.

By early April 1944, Bates, Crowell, and Cauthery had collated and analyzed the data from their wave observation network. A problem that bedeviled them was the variability and subjectivity of the human observers at the Coast Guard stations. No two people saw or reported quite the same thing. Oh, to have had wave heights accurately measured by impartial instruments. Furthermore, the observers reported both maximum and average wave heights in a three-minute period, and the forecasters needed to sort out what these meant in terms of what affected landing craft in the surf. It turned out that a value halfway between the average and the maximum was meaningful for operational forecasting of waves for landing craft. They called this value the "predicted height." After the war, as Sverdrup and Munk studied waves in more detail, they defined the term "significant wave height" as the average of the one-third highest waves in a three-minute period. For the purpose of deriving a practical technique for wartime wave forecasting, the numerical difference between significant wave height and predicted height was negligible.[13] Significant wave height is still used today in many applications of wave reporting and forecasting.

The observation stations along the East Anglian coast were exposed to waves approaching the coast over a wide range of angles. Bates used data from these stations to estimate how much refraction reduced wave heights. They had no idea from which direction waves would approach the Normandy beaches on D-Day, so they had to be ready to adjust the deepwater wave height predictions for the actual direction on that day.

A basic reference publication on English Channel weather, the *Weather Handbook for the English Channel* (published by the Air Ministry Meteorological Office), stated that wind waves in the channel depended in large degree on tidal currents in the channel.[14] This concerned them until they realized that currents along the invasion beaches ran parallel to the beach and therefore would not influence wave heights to any appreciable degree. This solved another of the section's objectives and completed their studies of shallow-water effects on waves.

Wave observations from the Coast Guard stations verified the wave forecasts they had made over the past several months, so they were satisfied with the accuracy of the Sverdrup-Munk method. Their final objective had been solved.

They developed graphs, which they called "surf prediction diagrams" (see Figure 4), for each beach with different characteristics: the British and Canadian beaches (Gold, Sword, Juno), and the American beaches (Utah, Omaha).

To begin a wave forecast, they got predicted wind speed and direction from the meteorologists. They estimated wind duration and entered their surf prediction diagrams with the wind data and came up with a predicted deepwater wave height. Finally, they applied the appropriate correction factor for the angle at which the waves would reach the beach and arrived at the surf forecast for each beach. By early April, the Swell Forecast Section was ready to forecast surf for the invasion, a mission actually assigned to the British Admiralty to whom they were seconded.[15]

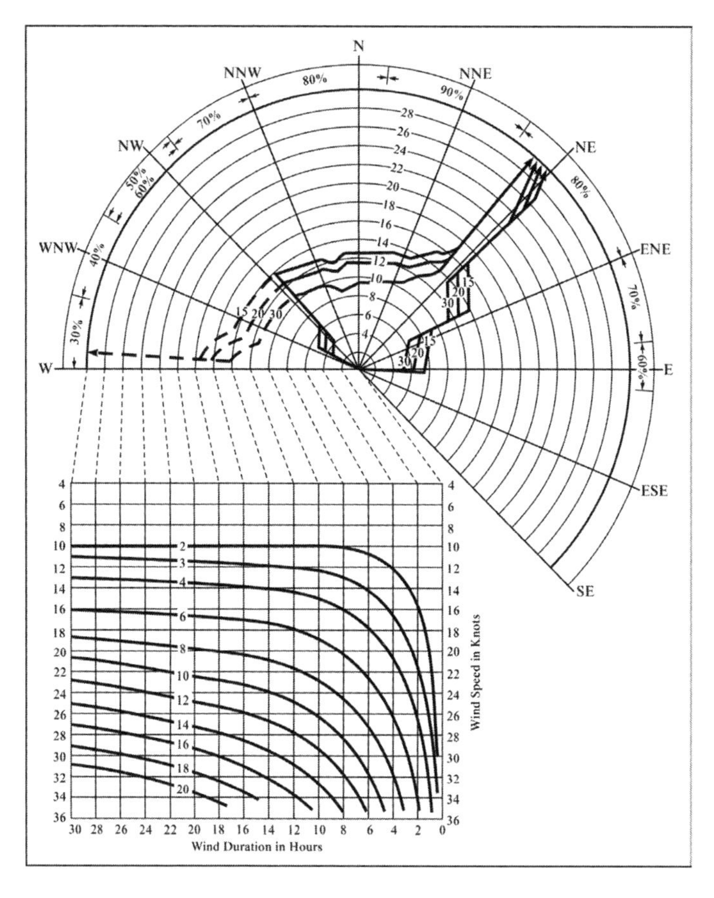

FIGURE 4
Surf prediction diagram for Omaha Beach prepared by the Swell Forecast Section one month before D-Day. Adapted from Crowell, *Surf Forecasting for Invasions during World War II*, 92.

MAJOR STORM SYSTEMS TRAVEL from west to east, from the United States across the Atlantic Ocean to Europe. To forecast weather along the French coast, it was necessary to know the weather patterns farther west: over North America, Greenland, Iceland, the Atlantic Ocean, and England. This gave the Allies an advantage over the Germans because England was their home base, and to a large extent, the Allies controlled the North Atlantic.

Weather stations in these countries reported meteorological data two to four times a day to central meteorological agencies in both the United States and

Britain. Ships at sea also reported weather observations, but their reports were not always effective because the ships were moving and frequently were not where they needed to be for the most useful weather observations. After all, ship captains try to avoid bad weather whenever they can. To provide additional upper-air information at sea, the British Royal Navy stationed two ships at fixed locations west of the British Isles. Their sole mission was to remain within small areas and report weather conditions from far offshore.

By the time the crucial D-Day weather forecast was made, the Royal Air Force was flying daily weather reconnaissance missions along tracks that extended from Gibraltar north to northern Norway and from these coasts several hundred miles to sea. They flew British long-range bombers at different altitudes to gather weather data at low, medium, and high altitudes and transmitted the information to the meteorological centers in England.

When the data from this network of monitoring stations reached the forecast centers, they were disseminated to the US Army Air Forces (code name Widewing) in England, to the British Admiralty, and to the British Air Ministry at Dunstable. These were the centers responsible for the operations of the British and US aircraft, naval vessels, and ground forces in Europe. They all needed accurate weather information and forecasts.

~~~

IN EARLY FEBRUARY 1944, General Harold Bull (SHAEF's deputy chief of staff for operations and Dr. Stagg's superior) asked Stagg to prepare practice D-Day forecasts to be delivered to Eisenhower's staff each Monday morning. The meteorologists were told to assume that the following Thursday would be D-Day. Dr. Stagg was told that the purpose of the practice forecasts was to "acquaint the members of the Supreme Commander's staff with the capabilities and limitations of the Meteorological Section."[16]

On March 6, Colonel Donald N. Yates replaced Colonel Tieman as Stagg's deputy and served in this position throughout Operation Overlord. Dr. Stagg was sorry to see Colonel Tieman go but welcomed Colonel Yates's more dynamic personality and greater knowledge of meteorology.

The American meteorologists at Widewing and the British at Dunstable conferred among themselves every Sunday night by telephone to discuss the
~~~

weather prospects for the following week. Then Stagg telephoned each forecast center and discussed the meteorological situation with them individually. From these discussions, he drafted his forecast for the next five days. Early on Monday morning he again consulted each of the weather centrals to learn of any changes that had occurred overnight that might alter his draft forecast. Then he finalized the forecast and gave it to General Bull for presentation to the supreme commander and his staff later that morning. After several weeks of these trial forecasts, the Admiralty forecasters and the Swell Forecast Section were brought into the discussions to add their expertise, including sea and swell forecasts, to the overall weather forecasts.

The Americans had a long-range forecast technique, often called the analogue method, with which they felt comfortable making five-day forecasts. Professor Irving Krick originated the method at Caltech. Now, as a US Army Air Forces lieutenant colonel and one of the two D-Day forecasters at Widewing, Krick aggressively advocated this concept. The two British weather centers thought the method had no merit.

Dr. Krick had collected more than forty years of daily surface weather charts of the Northern Hemisphere. To make a long-range forecast, an analyst looked through the charts until he found one that looked like the present day's chart. Then he simply assumed that the weather for the next five days would be the same as it had been for the five days following the matching chart. The biggest fallacy in this method, and the reason it was unreliable, was that Krick's charts were only surface charts at twenty-four-hour intervals. Circulation in the upper atmosphere drives the surface circulation, but without charts of the upper atmosphere, the analyst had no idea what the upper circulation had been at the time of his matching surface chart. The only times when Krick's method worked were when the upper-air circulation patterns just happened to be the same, but the analyst had no way of knowing this. Without upper-air information, the analogue method simply was unreliable.

The British meteorologists at Dunstable, being more familiar with forecasting for the British Isles, said it was folly, or worse, to even try to forecast weather over the English Channel more than twenty-four or thirty-six hours in advance. They did not like or trust Krick's method, and as it became obvious to Stagg before long, they did not like or trust Krick himself. The chief meteorologist at Dunstable, Dr. C. K. M. Douglas, threatened to take no part in making long-range forecasts until it was pointed out to him that if he did not, General Eisenhower would have no choice but to rely solely on the American forecasts utilizing Krick's analogue method. Dr. Douglas reluctantly agreed to try to provide the supreme commander some long-range "guidance" to counterbalance what he considered the unscientific and unreliable Krick analogue method.

Dr. Stagg also struggled with the question of five-day forecasts, for being familiar with British weather, he had little faith in the analogue method and considered stepping back from any long-range discussions. He realized that if he did, Colonel Yates (who in addition to being Stagg's deputy was also chief of the US weather services) would chair the discussions and might favor Dr. Krick's position. He came to the same conclusion as Dr. Douglas but kept his misgivings to himself and continued to work on a better way to prepare for the weekly briefings.

By early March, some of the military officers, particularly the Americans, chafed at working with a British civilian, so on April 18, 1944, Dr. Stagg became Group Captain Stagg, Royal Air Force, equal in rank to his American deputy.

These practice conferences continued until April 17, when Stagg was asked to present his five-day weather forecast at General Eisenhower's weekly conference with his entire staff. When Eisenhower introduced Stagg, he said, "I want to hear what our weather experts can do. Each Monday, until then [D-Day] Group Captain Stagg [apparently anticipating Dr. Stagg's promotion] will tell us what he thinks the weather will be for the rest of the week and on each following Monday, he will tell us how his forecast has worked out. We'll have a check on that part from our own experience." Stagg was "momentarily overawed" by the responsibility and all the high ranking officers present but pulled himself together and gave a credible presentation.[17] The following week, the general invited even more of his high-ranking staff to the conferences including "perhaps 25 to 30 admirals, generals and air marshals, add or subtract a rank or two," according to Stagg.[18]

To prepare for his meetings with General Eisenhower and his staff, Group Captain Stagg conferred with the three main weather centrals, which were spread over fifty miles. The United States Army Air Forces forecasters were located in western London, the British Admiralty meteorologists were in central London, and the civilian British Meteorological Office was thirty miles northwest of the city at Dunstable. The Swell Forecast Section was colocated with the Admiralty in London. Dr. Stagg was located at SHAEF at Bushy Park (King Henry VIII's former deer park) immediately adjacent to the US weather forecasters at the Widewing installation.

Forecasters from the air and naval commanders in chief were invited to join the discussions. These forecasters advised the operational units and on D-Day would provide weather information and forecasts to the units they supported. Stagg realized that it was imperative that they be a part of, and agree to, the Overlord forecasts. It would not do to have General Eisenhower decide to launch the invasion only to have one of his key players, the Army Air Forces for example, say the weather was unsuitable.

Since the participants were so widely spread out, the weather discussions were done by conference calls over a dedicated telephone network that linked all the offices. Each participant could talk and listen to all the others. The calls were electronically scrambled to preserve security, and Stagg likened the discussions to "disembodied voices around the invisible conference table."[19]

The American army forecasters at Widewing, the British civilian meteorologists at Dunstable, and the Royal Navy forecasters at the Admiralty conferred among themselves every Sunday night to try to arrive at a consensus about the weather conditions to be expected over the next five days. Then Group Captain Stagg telephoned each forecast center and discussed the meteorological situation with them individually. From this discussion, he constructed his forecast for the next five days and presented it to the supreme commander and his staff on Monday morning.

Each group of meteorologists worked from the same set of raw weather data, but the weather maps they drew from the data were not necessarily identical. In early 1944 there were no computers, FAX machines, or other means of rapidly transmitting graphic documents such as weather maps. Although facsimile machines were available to all the main weather units in Europe by the end of the war, not even SHAEF headquarters had them in April and May 1944. The coded raw data came into each center, usually by teletype, where it was decoded and plotted on base maps. Then meteorologists at each center drew in low and high pressure areas and barometric pressure contours and interpreted what it all meant in terms of weather. Although they started with the same raw data, differences in plotting and interpreting the data sometimes led to significant differences in the maps.

A representative of one of the centers chaired each teleconference, with the chairmanship rotating among the centers on a two-day basis. The meetings began with a description of the current weather map to assure that each group was using essentially the same map. Then the chairman would start the discussion by describing the weather situation as his center saw it, how it had evolved, and how they expected it to change in the next few days. If the three centers essentially agreed, Stagg's job was simple and he got them to agree on a consensus prognosis. This rarely happened, so Stagg asked the disagreeing center to state the reasons for its nonconcurrence. This usually launched a spirited, often heated discussion, which Stagg had to guide to an agreed forecast. He tried to steer the participants toward consensus by going from a broad overview of the weather to the daily details for the next five days. Stagg wrote, "It was admittedly an unorthodox procedure for producing a weather forecast but in the circumstances of the time, the state of the science in 1944, and the demands on us, none of the participants could suggest improvement."[20]

During the conference calls, two (usually) of the three oceanographers in the Swell Forecast Section sat in with the naval forecasters at the Admiralty and listened to the meteorological discussions. As soon as the general surface pressure gradients seemed certain, they prepared detailed wind fields for the channel, then turned to their surf prediction diagrams (see figure 4) to prepare wave and surf forecasts for the channel and the Normandy beaches. Their sea and swell predictions were added at the end of the meteorological prognosis to complete the weather forecast for the next five days.

Initially, the conference calls were held each Sunday evening. When they reached two hours in length, which was as often as not, Stagg halted them and insisted on agreement of a weather situation for the next five days. He used this to prepare a preliminary forecast. The conferees would meet again early on Monday morning to discuss any changes that had taken place overnight. Then Stagg would finalize the forecast and present it to General Eisenhower and his staff.

~~~

THE PRACTICE FORECASTS continued as the supreme commander and his staff planned the minutiae of Operation Overlord. There were plans upon plans and details upon details to be worked out. They expected Overlord to be a ninety-
~~~

day campaign from the landings at Normandy to the time the Allied armies reached the River Seine and liberated Paris. The amphibious landing phase at Normandy was named Operation Neptune and was a suboperation within the overall Operation Overlord.

On May 22, Group Captain Stagg was informed by General Bull that General Eisenhower had selected Monday, June 5, as D-Day. On May 29, General Eisenhower and his staff would move to the advance command post at Southwick House, a spacious mansion in Portsmouth, about seventy miles southwest of London. This would put the supreme commander and his staff in much closer proximity to the operational commanders and communications networks of the Royal Navy command center at Portsdown Hill, inland from the British naval base at Portsmouth. Stagg and Yates would move also and would share office space with Commander Fleming, the naval commander in chief, in a Nissen (Quonset) hut behind Southwick House. The space was cramped and crowded, hardly befitting work upon which so much depended. They shared sleeping quarters in a tent.

Stagg met with Eisenhower and his staff as usual on Monday morning, May 29. He reported that the forecasters were predicting generally favorable weather for the rest of the week, with some deterioration toward the weekend. Someone pressed Stagg for his opinion of what the weather would be like next Monday and Tuesday. This was a week away, and Stagg felt the conditions were too unsettled to forecast much past Thursday. He answered that both Monday (June 5) and Tuesday (June 6) could be stormy.

The conference calls continued during that last week before the invasion. At the Wednesday evening call, each of the forecast centers seemed at odds with the other two.[21] The Americans at Widewing thought a front would move through on Saturday with good weather behind it and said that D-Day could go on June 5 as planned. The British forecasters at Admiralty and Dunstable thought the front would linger over the channel with very unfavorable weather on June 5. Stagg felt that the Americans were too optimistic and sent a message to General Eisenhower that in his view, the prospects for launching the invasion were poor on Sunday (June 4) and Monday (June 5) and probably on Tuesday (June 6) as well.

He hoped the situation would become more clear during the teleconference Thursday afternoon, but the centers had not changed their opinions, and if anything, Lieutenant Colonel Krick at Widewing was even more optimistic than he had been before. Stagg tried hard to understand how experienced forecasters at the two centers, looking at essentially the same maps, could come up with such different prognoses. He finally decided that Krick was so devoted to his analogue system that he could not accept any other explanation for weather development.

Dr. Sverre Petterssen, a senior meteorologist with Dr. Douglas at the British Meteorological Office at Dunstable relied on meteorological theory to relate flow in the upper atmosphere to movement of the surface low and high pressure areas. Stagg felt this was the preferable approach and virtually ignored Krick's predictions after that. Time was running out, and Stagg and Yates were deeply troubled by the divergent positions.

When Stagg looked at the first charts early Friday morning, June 2, all hope of reconciliation vanished: a chain of lows was approaching from the Atlantic, and any of them could halt Overlord. The British forecasters at Dunstable were "unmitigatedly pessimistic," while the Americans at Widewing thought the weather on June 5 would be most favorable.[22]

Beginning on Friday, General Eisenhower started meeting twice a day with his staff. Group Captain Stagg told the generals and admirals that the whole situation was "full of menace," with high winds and complete overcast through at least Tuesday (June 6). The invasion was still on, though, and ships would start sailing within hours, into what he feared would be a disaster.

Stagg gave the generals the bad news at the Saturday evening briefing. He told them to expect high winds and seas, overcast skies, low visibility, and fog until Monday. General Eisenhower decided to delay the operation on a day-to-day basis. Ironically, the weather over southern England that night was clear and windless.

Stagg's Sunday morning (June 4) forecast called for all weather factors to be unacceptable the next day. The weather around London was clear and beautiful that morning. The navy and army commanders thought they could go ahead and were against delay. The air commander in chief, Air Chief Marshal Leigh Mallory, said his bombers could not carry out their missions in those conditions. That sealed the deal. General Eisenhower said, "In that case gentlemen, it looks to me as if we must confirm the provisional decision [to delay] we took at the last meeting."[23] He gave orders to recall all ships that had already sailed.

When the charts were updated about 1300 hours, some changes gave Stagg a faint ray of hope. A depression far out in the Atlantic, which they had expected to reach the British Isles the next day and bring stormy weather Monday night, was slowing down and looked like it would not reach the English Channel until late Tuesday night. This meant that between the front that was about to hit them at that time and the depression in the Atlantic, there could be an interlude of good weather on Tuesday (June 6) just long enough to begin the assault.

Stagg expected the Sunday afternoon teleconference to go smoothly when everyone saw this window of opportunity in the charts. Wrong! Petterssen and Krick continued their arguments and held their positions.

After another weather conference that evening, Stagg went in to face General Eisenhower and his staff at 2130 hours.[24] He told them that a cold front had pushed through more quickly and farther south than had been expected and would pass through the channel area that night or early the next morning (June 5), which should bring improved weather Monday afternoon into Tuesday. He forecasted good weather through most of Tuesday, deteriorating some Tuesday night and Wednesday. He thought wind speed along the French coast would be Force 2–4 (4.6–18 mph), increasing from west to east. Stagg assured them that clouds would not be a problem. Wind waves and swell would be no problem either. Waves along the Normandy beaches would be three to four feet, except two to three feet on Utah Beach, which was shielded by the Cotentin Peninsula.

Stagg and Yates were excused but waited outside the conference room for the decision. General Eisenhower came out. He had just issued provisional instructions to launch D-Day at 0630 Tuesday morning, June 6. He smiled at them and said, "Well, Stagg, we're putting it on again; for heaven's sake hold the weather to what you told us and don't bring any more bad news." The ships and airplanes would soon be on their way, but the general wanted to hold one last review meeting, and the weather forecast would be the primary topic of interest.[25]

At 0300 Monday morning, June 5, Stagg conducted another teleconference with his forecast centers. Krick and Petterssen continued their bickering, but Stagg gained sufficient concurrence for his final weather forecast before he rushed off to General Eisenhower's last meeting before the bombs started dropping. A copy of the final wave and swell forecast, which was prepared by Lieutenant John C. Crowell of the Swell Forecast Section at 2200 hours on June 4, is included in table 1. It was attached to Group Captain Stagg's final forecast.

TABLE 1. SHAEF 5-Day Wave Forecast for 5–9 June 1944

Swell: In western approaches to English Channel, and south of 50 degrees N up Channel as far as the Cherbourg Peninsula: 6 to 7 feet Monday, decreasing to 4 to 5 feet Tuesday, 3 to 4 feet remainder of period, westerly direction throughout.

Sea: Monday, 5 June:
(a) Western approaches to English Channel 8–10 feet mixed sea and swell.
(b) Near the English Coast, in the Channel: 3–4 feet west of Portland Bill, 2–3 feet in the east.
(c) French Coast (except western Cherbourg Peninsula: 5–6 feet decreasing to 3–4 feet).
(d) Southernmost North Sea: 5–7 feet.

Tuesday, 6 June, D-Day, Areas as above.
(a) 3–4 feet wind waves.
(b) 2–3 feet becoming 3–4 feet in the west.
(c) And (d) 3–4 feet except for 2–3 feet in southwestern Bay of Seine.

Wednesday to Friday, 7, 8, and 9 June.
(a) 5–7 feet mixed sea and swell.
(b) 2–3 feet, risk of 4 feet.
(c) 3–5 feet, but 2–4 feet in Bay of Seine.

Note: Prepared at 2200 DBST, Sunday, 4 June 1944, by Lieutenant John Crowell, A.C.

Sources: Charles C. Bates, "Sea, Swell and Surf Forecasting for D-Day and Beyond: The Anglo-American Effort, 1943–1945" (2010), 16; and email to author, 7 February 2010; John C. Crowell, *Surf Forecasting for Invasions during World War II* (Santa Cruz, CA: Marty Magic Books, 2010), 87.

Group Captain Stagg told the assembled staff officers that the weather conditions had changed little in the few hours since his last forecast. He predicted low amounts of cloud and wind on Tuesday west-northwest to west, Force 3–4 (15–18 mph); waves on the Normandy beaches would be two to four feet. Wednesday, winds Force 3–4; Thursday-Saturday, winds Force 3–4, sometimes 5. Cloud cover after Tuesday would sometimes be complete, but the main requirements for low cloud cover would have passed. By then over 150,000 troops would have landed and should have established a solid beachhead. The tension that had been palpable in the room evaporated.

Stagg and Yates were dismissed but waited outside the conference room. Soon the generals and admirals emerged from the Southwick House library and told them General Eisenhower had made the final and irrevocable decision to launch Operation Overlord the following morning, June 6, 1944, at 0630 hours. It was now very early morning, June 5, and all they or anyone else on the planning staff could do was wait to see what the weather would be and how the landings would proceed.

As Monday wore on, their forecast proved correct: strong winds along the French coast overnight and into Monday afternoon produced waves that would have made amphibious landings just about impossible. A continuous overcast covered the area and would have made aerial and naval bombardments ineffective and undoubtedly would have interfered with the airborne and glider operations. Soon after midnight Monday, with the paratroopers already on their way, reports from the channel and Normandy confirmed that clouds were breaking up over the French coast. Winds along the beaches that morning were from the west to northwest at up to Force 4 and caused rough seas that hampered transferring soldiers to the landing craft, but the landings proceeded with relatively few weather-related mishaps.

In a report to the engineer's regional control officer, Lieutenant Donald Pritchard compiled a table of the actual wind and wave conditions reported at the Normandy beaches on D-Day, June 6 (table 2).

Wave heights at 0300 hours at Utah and Omaha Beaches, ten nautical miles offshore in the area where troops transferred to the landing craft, were three to four feet and occasionally up to six feet. Wave heights at Utah and Omaha Beaches were two and three to four feet, respectively, at 0640 hours. The wind that morning was from the northwest. The Cotentin Peninsula sheltered Utah Beach more than it did Omaha Beach, and in oceanographic terms, gave Omaha Beach a longer wind fetch and consequently higher waves. By noon the surf had decreased to two feet or less on both American beaches.

Note also from Pritchard's table that twenty-eight out of fifty-seven skirted Sherman tanks launched at Omaha and Utah Beaches were swamped by the waves and sank. These tanks, known as duplex drive tanks, had waterproof canvas screens attached to their hulls that were folded down onto the upper part

TABLE 2. Observed Wave Conditions at NEPTUNE Beachheads, 6–7 June 1944

DATE	OBSERVATION AND ITS LOCATION	
June 1944 (D-Day)	OMAHA	Troop unloading area 10 nautical miles offshore experiences gusty northwesterly winds of 12–16 knots. Wave heights of 3–4 feet with occasional interference waves up to 6 feet. Choppiness makes personnel transfer difficult.
0540–0640 DBST	ALL BEACHES	Skirted Sherman tanks (DD-Dual Drive, treads and propellers) launched even though operational limit is 1-foot high waves. *Consequences:* UTAH: Launched 0.6 miles offshore into 2-foot waves. 27 out of 28 tanks reach shore. OMAHA: Launched 3.5 miles offshore into 3–4 foot waves. 27 out of 29 tanks sink. GOLD & JUNO: Launched 0.4 miles offshore into 3-foot waves. 42 out of 58 tanks reached beach. SWORD: Launched 2.2 miles offshore into waves less than 2 feet high. 24 out of 24 tanks reach beach.
1200 DBST	UTAH OMAHA	Surf less than 2 feet high. Transport unloading area continues with choppy waves 3 to 4 feet high; surf 2 feet high.
1800 DBST	OMAHA	Surf 1 to 2 feet high; offshore waves 2–3 feet high. Wind remains northwesterly 12–16 knots.
7 June 1944 (D-plus-One) 1200 DBST	OMAHA	Offshore waves still 2–3 feet high . . . northwesterly wind speed of 10 knots or less. Surf 1–2 feet high.

Notes: Memorandum to Regional Control Officer, 21st Weather Squadron dated 30 June 1944. Fletcher, David. *Swimming Shermans: Sherman DD Amphibious Tank of World War II.* Oxford: Osprey Publishing, 2006; Pritchard, D.W., 1st Lt., AC.

Source: Charles C. Bates, "Sea, Swell and Surf Forecasting for D-Day and Beyond: The Anglo-American Effort, 1943–1945" (2010), 18.

of the hull when not in use. When the tank needed to "swim," compressed air piped into vertical rubber tubes erected the screens. The screens rose higher than the top of the tank's turret and provided about three feet of freeboard. A pair of propellers at the rear of the duplex drive tanks propelled them in the water. The duplex drive tanks had been tested only in inland waters and relatively calm offshore sea conditions, and the rough seas encountered on D-Day were too much for many of them. Also, many were launched much farther from shore than was planned, and this contributed to the high number of duplex drive tanks lost on D-Day.

What had conditions been on Monday, and what would the soldiers and sailors have faced if General Eisenhower had not delayed Operation Overlord one day? Gale-force winds (up to 35 mph), high seas, complete overcast, and low visibility prevailed. These conditions would have severely hampered Overlord, if not turned it into a disaster.

Incredibly, forty years later, Dr. Krick still maintained that on June 5 the weather had been acceptable for Overlord. At a symposium convened by the American Meteorological Society and attended by many of the people involved in the D-Day forecast, including Krick and Charles Bates, Krick said, "But they could have gone on the fifth, the data that we have from all of the maps that were drawn subsequently showed that the front had been clear over in Paris by the morning of the fifth and the winds in the Channel actually were less than they were at times on the sixth."[26] He went on to plug his analogue method of long-range forecasting, which had been repeatedly discredited by then. Charles Bates, who attended Dr. Krick's presentation at the symposium, said that Krick's statement "drove me nuts."[27]

What were the Germans doing all this time? Captured documents and interviews with German officers after D-Day showed that the German Army had been on maximum alert all through the relatively quiet days of May but had relaxed when the weather in early June became so unsettled.[28] Some of their channel defense forces were withdrawn for exercises inland. General Rommel left his headquarters in Paris for Berlin to celebrate his wife's birthday and to confer with Hitler. On the morning of June 5, the German chief meteorologist, Dr. Hans Muller, sent a forecast to the Luftwaffe's chief meteorologist in Paris showing the same break in the weather that Stagg saw, but the high commanders were not there to see it.[29] Hitler liked to sleep late and had left standing orders not to wake him early, so he wasn't even informed of the invasion until he awoke in midmorning.

Would weather conditions have been more favorable for the invasion later in June? June 7 was not as good as June 6, and if they had not gone then, the next favorable combination of tide and daylight (but not moonlight) would have been June 17 or 18. On June 18 a cold front advanced over the British Isles, and a depression in the Mediterranean moved into France. This spawned the most intense summer storm in forty years, which ravaged the channel June 19–21. It was so intense it garnered a name: the Big Storm. If troops had been trying to land at Normandy then, the results would have been unthinkable for the Allies. After seeing the effects of the Big Storm, which interrupted landing operations for days, wrecked many ships and landing craft, and destroyed one Mulberry artificial harbor, General Eisenhower sent Group Captain Stagg a note saying, "I thank the gods of war we went when we did."[30]

With the possible exception of a few days in early August, June 6 was the best day that entire summer to have launched Operation Overlord / Operation Neptune.[31]

4

SURF FORECASTING FOR RESUPPLY AFTER D-DAY

IN MID-MAY 1944, THE OFFICERS and enlisted men of the 21st Weather Squadron's Station WZ at Woolacombe received orders to report to a staging area in Wales.[1] This could mean only one thing: the long-awaited invasion was imminent. Bob Reid and hundreds of other soldiers boarded Liberty Ship 234 on June 1 for the voyage to an as-yet undisclosed landing beach. Security was so strict that until after the ships sailed, only the highest ranking officers and those who needed to know were told where the landings would take place. The American slogan "Loose lips sink ships" was taken very seriously.[2] Limiting the number of people who knew their destination reduced the likelihood that the Allied plans would reach German ears.

Liberty Ship 234 remained at anchor in Bristol Bay for four days as a large fleet assembled there. On June 5 the ship weighed anchor for Normandy.

Bob Reid was assigned to a weather detachment that would operate after the invasion beachhead was established. He now knew where they were going and briefed his men on their destination and plans.

High winds, rain, and rough seas battered them as soon as they reached open water. This was the heavy weather the meteorologists had forecast for the original D-Day date, and the men on Liberty Ship 234 were experiencing it firsthand.

At breakfast on June 6, the troops in the armada learned that the first wave of landing craft had hit the beaches at Normandy at 0600 hours and had established a beachhead. That afternoon Liberty Ship 234 and its fleet rendezvoused with other convoys off the Isle of Wight, then crossed the English Channel after midnight and anchored off Omaha Beach at 0800 hours on June 7.

For the next two days, assault troops and supplies streamed ashore as the navy fired shells of every size and description over their heads at German positions inland. The cloud cover that was so dreaded by the Air Forces cleared. Allied bombers dropped everything in their arsenals on the German ground forces while fighter planes provided a defensive shield against the Luftwaffe.

By June 9, the Allies (Americans, British, and Canadians) had pushed the Germans far enough inland for noncombat personnel like Bob Reid and his weather detachment to go ashore. After a last supper aboard the 234, Reid and two enlisted men joined their jeep aboard a landing craft tank for the short trip to Omaha Beach. Upon reaching shore at 2000 hours, the bow ramp of the landing craft tank was lowered, and the team simply drove into water two feet deep. They dodged the detritus of war littering the beach and made it safely to a transit (rendezvous) area to spend their first night ashore. They did not dare attempt to find other members of their detachment until daylight because they had no idea where the front line was.

The transit area was nestled in a hedgerow, and Reid and his team spent their first night at Normandy in foxholes, sleeping on top of their gear. They kept their helmets on and weapons close at hand. When German planes flew over near midnight, the Americans were treated to a spectacular fireworks display as antiaircraft guns in the adjacent field opened fire on the planes with tracers.

The next morning they found their detachment area, and Don Pritchard arrived in late afternoon to take command. He was accompanied by Major Harry Richard Seiwell. Seiwell was now assigned to the 5th Engineering Brigade at Normandy to assist in constructing and operating an artificial "Mulberry" harbor. Since the Germans had destroyed all the French ports, the plan was to unload supplies at the artificial harbor until permanent dock facilities could be rebuilt along the coast. This operation was highly dependent on waves, tides, and other beach conditions, so good forecasts were mandatory. Major Seiwell wanted to have Reid and Pritchard temporarily assigned to the engineering brigade to forecast swell and surf and to survey the bottom at Omaha and Utah Beaches. The surveys would be used to evaluate the assumptions about beach conditions that were made during the preinvasion planning and ultimately to improve planning for future amphibious landings.

Reid and Pritchard asked Major Seiwell if he had authorization from Colonel Moorman, their commander, to assign them to the engineering brigade. Although Major Seiwell was a competent scientist who had earned the respect of Dr. Sverdrup when they worked together to establish the wave forecasting program for Operation Torch, his managerial style did not always follow the strictest military or ethical protocols, and he left them wondering if he had spoken to Colonel Moorman at all. On June 13, Major Seiwell returned and told them they were temporarily assigned to the Engineer Command at Omaha Beach, code name Mars. Although Reid and Pritchard still were unsure of his authority, the mission was clearly within Colonel Moorman's directions to them, and the engineering brigade was a logical user of their services, so they acquiesced to the major's directive. They were to provide weather and surf forecasts to the Mars G-2 (intelligence) section and to work with the Engineer Survey Group on a bottom survey of Omaha Beach. They worked out of Weather Detachment M, headquartered at a fighter aircraft field near the hundred-foot-high cliffs at Grandcamp.

Reid and Pritchard were well advised to question Seiwell's directive assigning them to the brigade, and the fact that they did not receive a clear answer indicates that the major may have been cutting corners in military procedures. Major Seiwell seemed overly ambitious to them, and they suspected he was trying to carve out a niche for himself in beach intelligence in the war effort. It is likely that they were right because in November 1944 he succeeded in establishing a large beach intelligence organization in Paris to which he got Reid and Pritchard assigned. Major Seiwell also tried to get Charles Bates and John Crowell, who by then were captains, assigned to his Paris organization, but they were already scheduled to join the British commander in chief, Headquarters, East Indies Command, in Colombo, Ceylon, effective January 1945.[3] Thus they avoided being caught up in Major Seiwell's "empire" as Charles Bates called it.

Early on the morning of June 15, Reid and Pritchard left camp to start their surf and weather observations along the cliffs at Pointe du Hoc, a hundred-foot-high promontory at the boundary between Utah Beach on the west and Omaha Beach on the east. Their forecasting mission could have ended before they took their first observations when Prichard's boot legging caught in a wire strung along the edge of the cliff. A small explosion nearby was followed several seconds later by an enormous explosion on the beach. At first, they thought they were being shelled by friendly fire from the battleship *Texas* that lay offshore, but after checking for survivors and finding no one was hurt, they realized Pritchard had snagged a booby trap trip wire. The first explosion, a propellant charge, hurled an eight-inch artillery shell several hundred yards to explode harmlessly on the beach. The incident even had a beneficial side effect. They had been constipated for several days but, after the explosion, no longer were.[4]

After regaining their composure and attending to their hygiene, Reid and Pritchard began their wave observations by scanning the surf zone. It was difficult to estimate wave heights from such a vantage point, even with binoculars, and their accuracy undoubtedly was lacking, but they did their best and sent their observations to Colonel Moorman at 21st Weather Squadron Headquarters and to Major Seiwell at the 5th Engineer Brigade. To improve their efficiency, Pritchard decided that he would make the weather and wave forecasts and assigned Reid the task of making the beach surveys.

The primary purpose of the surveys was to compare mean sea level and actual beach profiles (water depths versus distance offshore) with the information used earlier in the planning phase of Operation Overlord. The planners had relied on tide information available at the time, but they were unsure of the accuracy of the existing charts. Bathymetric charts provide water depths referenced to tidal datum, such as lowest astronomical tide, mean sea level, or any of several others. It was important that the engineers learn more about Normandy tides, sea levels, and beach profiles so they could improve their methods for future mission planning.

~~~

TIDES ARE ONE OF THE ENGINES that keep the ocean running, especially near shore. Strictly speaking, tide is the vertical movement of the water's surface due to the gravitational attraction of the moon and sun on the sea.[5] As the surface rises or falls, the water has to go somewhere to compensate for the deepening or shoaling of the water. This creates tidal currents.

Water rushes through inlets and alternately fills and flushes bays and estuaries. This is important not only to the life cycles of the fauna and flora that live in these places but also to the lives of the bodies of water themselves. Many species of ocean life spawn in estuaries and depend on the currents to carry their
~~~

eggs and larvae into the ocean to mature. Other creatures spawn in the ocean and trust the tidal currents to carry their progeny into the bays. Passes, inlets, and bay bottoms are scoured of silt that would eventually choke and fill these nearshore bodies of water were there no currents. Pollutants and other forms of human waste are flushed from nearshore regions by the tides. The effects of this trash on the deep ocean are another story, however.

As the water level drops from one high tide to the next low tide (or rises from low tide to high tide), the vertical distance through which the surface travels is called the tide range. The range varies tremendously from place to place on the globe and from time to time at any given location. The very highest tides on earth occur in the Minas Basin at the northeast end of the Bay of Fundy, Nova Scotia.

According to the website of the town of Burntcoat, Nova Scotia, the *Guinness Book of World Records* listed Burntcoat Head as having the world's highest recorded tide range.[6] The highest average tide range at Burntcoat is 47.5 feet, with an extreme range of 53.6 feet.

Such extreme tidal ranges are rare and in fact are found only in the Bay of Fundy and Ungava Bay, Quebec. Tides of only somewhat lesser range (on the order of thirty-three to forty-six feet) are found in a number of places, though, such as Inchon, Korea; the Sea of Okhotsk, Russia; the northern coast of Australia; Anchorage, Alaska; at the head of Cook Inlet in the United States; the Bristol Channel of England; and Mont St. Michel and the Normandy coast of France.

Gravity is the force that drives tides on earth. As the moon revolves around the earth, the waters of earth are attracted toward it. The sun also attracts earth's waters as we make our yearly trek around the sun, with the planet spinning on its axis every twenty-four hours. The gravitational attraction between two bodies depends on their masses and on the distance between them: the larger the bodies, the larger the gravitational attraction; the closer the bodies are to each other, the stronger the attraction between them. The sun is about twenty-seven million times more massive than the moon, but it is nearly 390 times farther away from earth. When all the math is done, the moon's effect on earth's waters is a little more than twice that of the sun. Thus, the moon is the primary tide-producer on earth, although the sun does make significant contributions to our tides.

In Figure 5, we are looking down on the north pole of a spherical earth that is covered with a uniformly deep ocean (if there were no moon or sun), with no continents or other land masses. It is easy to imagine the moon's gravitational

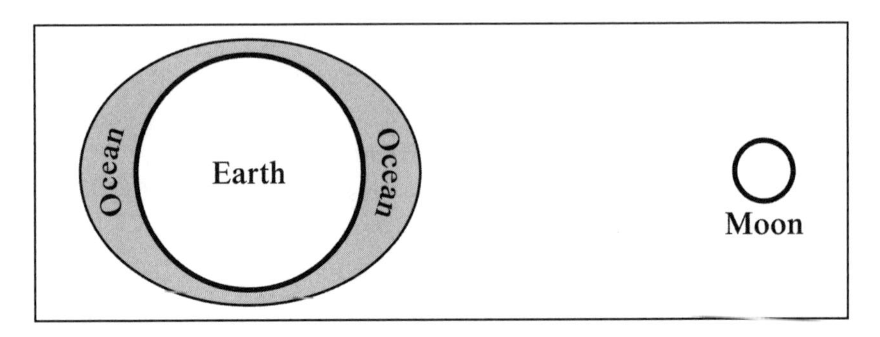

FIGURE 5
Looking down on the north pole of a spherical earth that is covered with a uniformly deep ocean. The moon causes the ocean to bulge outward in this most simplistic representation of the earth-moon system.

attraction causing the water to bulge toward the moon on the side of earth facing the moon. The water is deeper on that side of earth, but why is there also a bulge on the other side?

To answer this question, we must look at the earth and moon as a combined system of bodies rotating together through space.[7] Think of having a baseball (earth) and a golf ball (the moon) tied together with a piece of string (gravity). Toss them into the air and they will rotate around a point somewhere between the two balls. The center of rotation will be closer to the heavier ball (the baseball, earth) than to the lighter one (golf ball, the moon). The same applies to the earth/moon system. They rotate around a point approximately 2,900 miles from the center of the earth (about three-quarters of earth's radius) on the side of earth facing the moon and along a line joining their centers. (This is in addition to the earth's daily revolution around its own axis.) This off-center rotation causes centrifugal forces in addition to the centrifugal force caused by the earth revolving around its own axis. Everything on the earth, including the ocean's waters, experiences these forces. The centrifugal force is everywhere directed outward from the earth and is stronger than the moon's gravitational attraction. On the side of earth facing the moon, the moon's gravitational attraction adds to the centrifugal force and causes the water to bulge toward the moon. On the opposite side of earth, the forces act in opposite directions, but since the outward-directed centrifugal force is greater than the moon's gravitational attraction inward, the ocean also bulges outward on that side of earth.

The location of the tidal bulge relative to any fixed point on earth is constantly changing because the earth rotates on its axis once every twenty-four hours. Thus if we stand in the ocean at the point directly beneath the moon, we will be in the bulge of water, and the water will be deep (high tide). Six hours later, our globe (with us standing in the same spot) will have rotated one-quarter of a revolution, and we will be standing at one of the points where the water is shallowest (low tide). Six hours later, we will be opposite the moon, and will again be in deep water. Six hours after that, we will be in shallow water, and finally in another six hours we will be back in the bulge of deep water under the moon. Twenty-four hours have elapsed, and we have been through two complete tidal cycles.

This two-dimensional figure (see Figure 5) might lead one to believe that the tide-producing forces are applied only along the equator, but in fact they act in three dimensions on every drop of water covering the entire globe. Water from everywhere on earth flows either toward the point directly beneath the moon or toward a point on the opposite side of the earth, creating the bulge there.

The sun's gravity attracts earth's waters just as the moon does, but with only about half the force of the moon. The sun's position relative to the earth and moon is constantly changing, so even though its tide-producing force is only about half that of the moon, it still has a significant effect on earth's tides. This is one of the reasons that tides at any given location change throughout the month and year.

Over the period of one month, the earth, moon, and sun vary in position as shown in Figure 6. In the first (upper) and third figures, the gravitational forces of the sun and moon are aligned with one another. This means that the tide-producing forces of the sun and moon are acting together to create especially high tide ranges. These conditions occur twice each month, at full moon (first figure) and at new moon (third figure) to produce "spring tides."

The name has nothing to do with the spring season of the year, since it happens twice every month. Think of it as the time of month when the tides "spring up" to their highest levels. When the moon is one-quarter (second figure) and three-quarters (fourth figure) of the way around the earth, the gravitational forces of the moon and sun are acting at right angles to one another. At these times of the month, the tide range is at its lowest, a condition known as neap tide. At times of the month between the four conditions shown in the figure, tides are at intermediate levels.

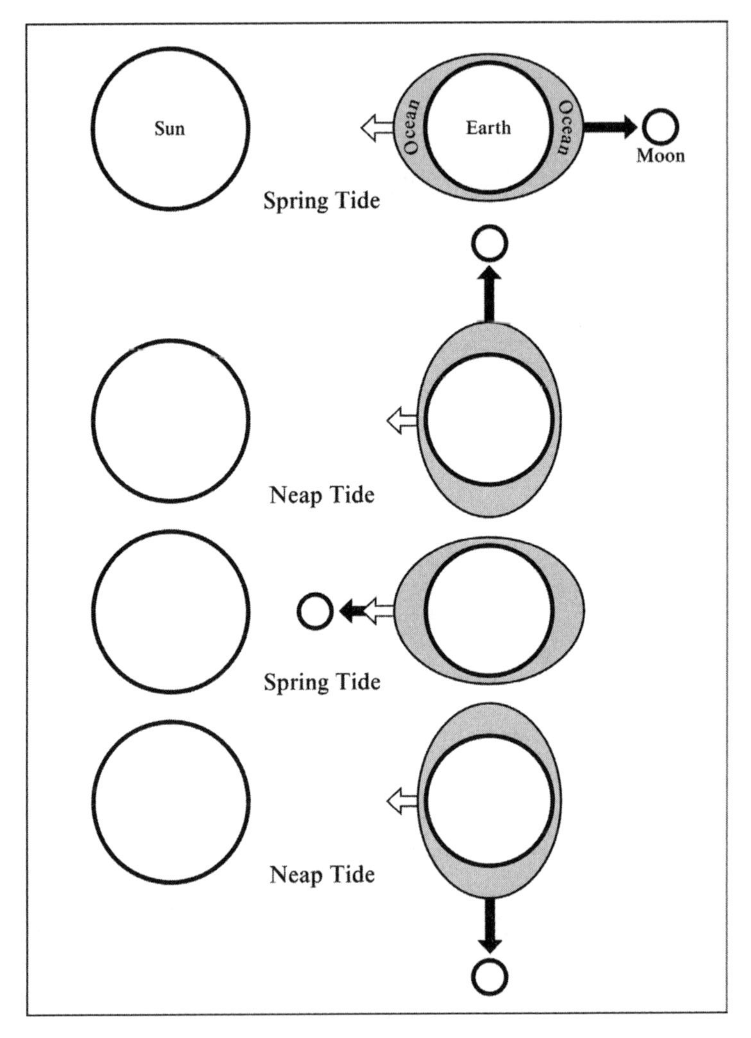

FIGURE 6
Effect of tide-producing forces of moon and sun (arrows) on earth's ocean.

There are many factors other than the relative positions of the moon and sun that affect tide levels, even on our hypothetical water-covered globe. Twice each month the moon crosses the earth's equator. Thus the bulge in the ocean's water shifts north and south of the equator twice each month. This causes the tides at any location to vary from day to day over the month.

The earth rotates counterclockwise (when viewed looking down on the north pole) on its axis once every twenty-four hours. This gives us our twenty-four-hour days. The moon rotates around the earth in the same direction, so in one day, a point on the earth rotates 360°, but it must rotate an additional distance (12.2°) in order to "catch up" with the moon. Therefore, high tide at that point is later by fifty minutes each day (the time it takes the earth to rotate 12.2°) because of playing "catch-up."

These are the principal astronomical factors that govern tides on earth, although there are many more that play lesser roles in the variation of tide height at every location on earth. There are so many variables affecting the tides that it takes 18.5 years for tides to exactly repeat at any location. However, since the forces that produce tides are astronomical bodies whose orbits are repeated and very well known, it is possible to predict tides anywhere on earth with extreme accuracy. Tide theory was well developed at the outbreak of World War II and has not advanced much since then. What has changed, though, is the ability to process enormous amounts of data with modern digital computers and crank out tide predictions everywhere on earth. Not only is it possible to forecast tides at any future date, but it also is possible to look backward in time and "hindcast" tides anywhere, at any time in history.

When we introduce real land masses into our water-covered globe, we deal with an ocean that is divided into basins of widely varying sizes, depths, and shapes. Some are virtually unlimited in extent, such as the Pacific and Atlantic Oceans, some are nearly enclosed such as the Gulf of Mexico and the Mediterranean Sea, and some are bordered by very complicated coastlines and estuaries. Some coasts slope gently from the shore seaward, while others have very steep slopes. All of these factors alter the ideal situation and create different tide characteristics in each place.

Daily tide ranges vary from virtually zero to over fifty feet. Most places on earth have two tides per day (semidiurnal), but some have only one (diurnal). In some places the two tides are of nearly the same height, but in others they can be quite different.

Figure 7 shows the tide range (in meters) at Normandy for the month of June 1944. The water levels are referenced to mean sea level, which is shown as zero on the figure. This is a semidiurnal tide with two nearly equal tides each day. Spring tides occur on June 7 and 21 and neap tides occur on June 15 and 30.

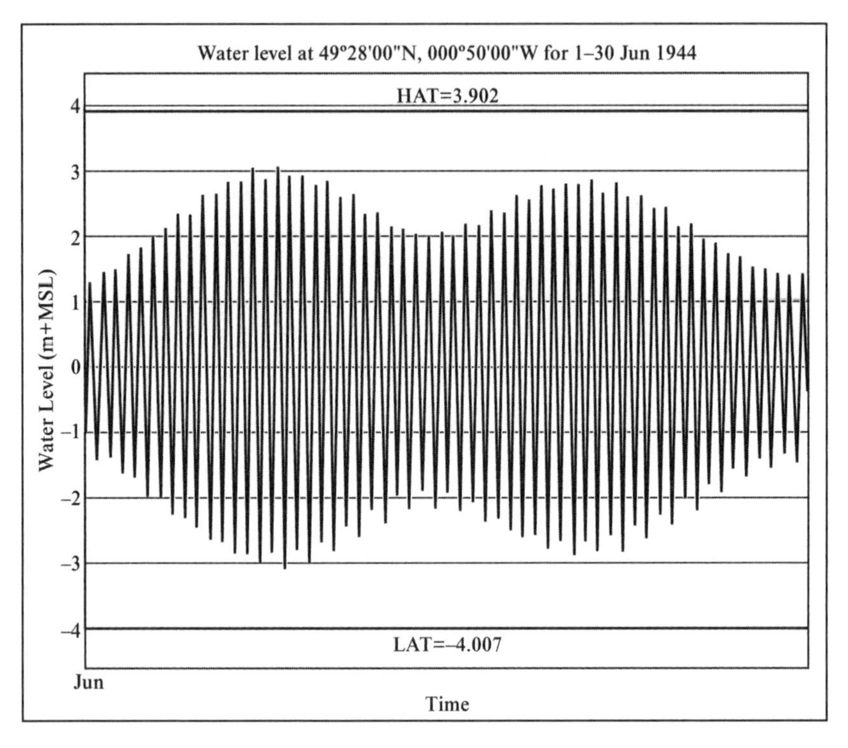

FIGURE 7

Normandy tides for June 1944: semidiurnal tides with two nearly equal tides each day. Spring tides (June 7, 21) and neap tides (June 15, 30) are clearly apparent. Courtesy Joris de Vroom, metocean consultant and maritime forecaster, BMT ARGOSS, Spacelab 45, 3824 MR, Amersfoort, the Netherlands.

The highest astronomical tide and the lowest astronomical tide are also shown. These are the highest and lowest tides that will ever occur at this location off the Normandy beaches of France.

Figure 8 shows the tide range for Normandy with an expanded time scale for June 5–8, 1944. Note that a low spring tide occurs in the early morning hours of June 6.

Since the forces that produce tides are celestial bodies whose orbits had been precisely known for at least a century, it was possible to forecast tides anywhere on earth with great accuracy. Tide theory was well developed at the outbreak of World War II, but actually using the theory to forecast tides (before the

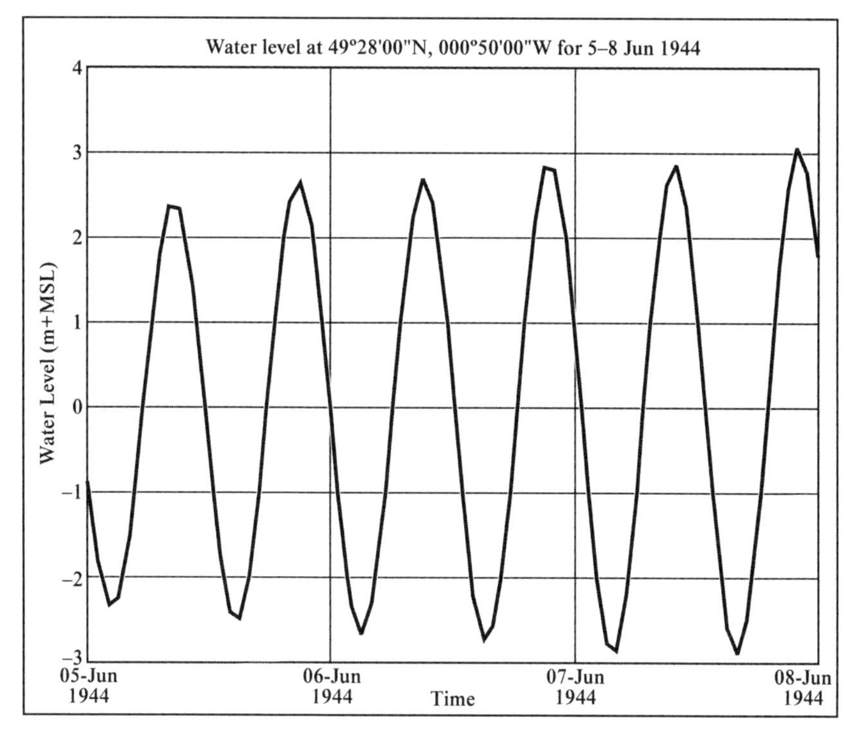

FIGURE 8

The tide range for Normandy with an expanded time scale for June 5–8, 1944. Note that a low spring tide occurred in the early morning hours of June 6. Courtesy Joris de Vroom, metocean consultant and maritime forecaster, BMT ARGOSS, Spacelab 45, 3824 MR, Amersfoort, the Netherlands.

invention of modern digital computers) was a time-consuming, labor-intensive endeavor. The process started by measuring water levels continuously for several weeks (preferably months) at the site of interest. These data were sent to a laboratory where analysts pored over the numbers to filter out variations in water level due to waves and wind. Then the analysts began the laborious mathematical calculations to extract the tidal constituents (pairs of numbers) from the time series measurements. The tidal constituent numbers then were used to hand calculate tide tables to predict tide levels (water depths and times of occurrence) at the site of interest. This procedure was repeated, including sending measurement crews into the field (sometimes in enemy territory) to measure at least two weeks of water depth information for every different location for which tide tables were required.

Much of the manual labor of hand calculating the tide tables could be avoided by using "tide machines" to calculate the hourly water levels. These large machines consisted of cranks, pulleys, and cables and occupied large amounts of space in research laboratories. However, with enough time, manpower, and long records of water levels measured at the site of interest to calculate tidal constituents, analysts could prepare accurate tide predictions, called tide tables. The tables forecasted the heights of high water and low water and their times of occurrence at places of interest. The process was cumbersome, labor-intensive, and time-consuming, but it worked. It required a dedicated machine for each site, with its gears and pulleys custom made to represent the tidal constituents of that site. A modern laptop computer using virtually the same theory can predict tides anywhere on earth, at any time in the future or past, in seconds. Figure 9 is a photograph of a tide machine used by the US government to produce tide tables for US waters from 1910 to 1965. Powered by a hand crank, it required that the gears, pulleys, etc. be sized and setup for each site for which tables were to be calculated, based on measured tide data from that site. A similar machine, built in 1872 for Lord Kelvin, a Scottish mathematician and scientist was one of two machines used by Arthur Doodson at the Liverpool Tidal Institute to predict tides at the Normandy beaches for the D-Day invasion.[8]

EARLY IN PLANNING FOR the D-Day landings, General Eisenhower's staff realized they needed to know the characteristics of the wave environment offshore at the Normandy beaches. The army turned to Dr. Garbis H. Keulegan, special assistant to the chief, Waterways Experiment Station Hydraulic Laboratory, US Army Corps of Engineers, to devise a means of remotely determining water depth. Dr. Keulegan suggested that it would be possible to determine wave characteristics from aerial photographs of a beach.[9] A light aircraft could be flown over the beach of interest at a known altitude and speed with a downward-looking camera taking overlapping photographs of the waves approaching the beach. An analyst would examine the photographs and note the point at which

FIGURE 9
NOAA's tide-predicting machine no. 2. This machine was used by the US government to predict tides from 1910 to 1965. It was powered by a hand crank (not shown) through the gears (*far left*). The gears, pulleys, chains, and other mechanical components of the machine were sized and arranged according to the tidal constituents at the location for which the tide predictions were being prepared. Doing so required measurements of tidal components at that location for a one-month period and reconfiguration of the machine for each new tide prediction location. Affectionately known as "Old Brass Brains," the machine is still maintained in working order by NOAA.

the incoming waves began to break. From bathymetric (depth) charts, the analyst would estimate the water depth at which the waves broke. This "chart" depth was only the water depth referenced to the chart datum, such as mean sea level. To obtain the actual water depth, the analyst had to adjust the chart depth for the variation in depth due to the tide at the time the photograph was taken. This adjustment was particularly important for the Normandy beaches because tides there can change the depth by as much as twenty feet.

Next, the analyst turned to a published tide table for the beach and determined the factor to apply to the chart depth to finally estimate the exact water depth at the time and place where the waves in the aerial photograph broke. Finally, using the oceanographers' rule of thumb that waves begin to break when they reach water depths of 1.25–1.5 times the wave height, the analyst could estimate the waves' heights. By knowing the plane's altitude and the optical characteristics of the camera, analysts could calculate the change in the lengths of waves over a known horizontal distance and estimate the water depth and the slope of the bottom.

With the Normandy beaches under Allied control after the invasion, General Eisenhower's planners wanted to assess the accuracy of the Keulegan method as used for the Normandy beaches prior to the D-Day landings. This request from Army Engineers Headquarters in London sent Lieutenant Bob Reid and his surveyors wading at Omaha Beach.[10] They needed to measure the bottom profile at ten-foot intervals along a line from the high-tide level seaward to low-tide level every three hundred feet along Omaha Beach. Reid got two five-man teams from the Mars Engineering Brigade assigned to their detachment to do the surveys.

The first step in the surveys was to set up what was called a "survey datum," a convenient fixed mark on shore that was visible all along the beach, to which their measured elevations could be referenced. They started the surveys by setting up a surveyor's transit on the beach, above the high-tide level. At low tide, a soldier (rodman in surveying parlance) would hold a surveyor's rod at the water's edge. A surveyor's rod is a long pole with marks at small increments along its length. When sighted through a level transit, the difference in elevation between the instrument's position and the rod's position is measured. A second soldier, known as an instrument man, would sight through the transit and read the depth of the bottom at that point. Then the rodman would move his rod ten feet shoreward, and the instrument man would take another reading with the transit. The procedure was repeated every ten feet up the beach until the rodman reached the high-tide level. That constituted one bottom profile. Then they moved three hundred feet farther along the beach and repeated the procedure. Having two crews cut the time in half, but the survey still took four days to complete.

The surveys were started on June 16, but the "Big Storm," the worst summer storm to strike the French coast in forty years hit Normandy and lasted through June 21.[11] After the storm passed, Lieutenant Bob Reid and his surveyors had to start over. The surveys were conducted near the time of spring tide so they could get as far out as possible at low tide.

All the measurements were referenced to the survey datum, but the reference datum that was really desired was mean sea level, which is the average water depth at a point, with all depth variations, mainly tides, averaged out. Thus the second part of Reid's task was to calculate water level variations at Omaha Beach during the times of their surveys.

They needed to take depth readings at a fixed point at Omaha Beach for two weeks. Bottom-mounted pressure gauges would have worked and are used for this today, but they were not able to procure any at Normandy in 1944, in the middle of a war. A telephone pole driven into the bottom somewhere seaward of the lowest tide level, along which they could have measured water level variations, would have worked, but they had no such pole. They improvised an array of shorter poles placed at intervals along a line from the highest water level to beyond the lowest. Each pole protruded about six feet above the bottom and had marks at one-foot intervals. With binoculars, the team observed the water level on the pole nearest the surf zone at one-hour intervals over a period of two weeks. They visually averaged out waves over several wave periods to try to arrive at a still-water level (the water surface, or depth, if there had been no waves). At the end of the two-week period, they had an hourly record of sea level at Omaha Beach, from which they could calculate mean sea level.

In order to check their measurements and calculations, they asked the engineers to dig a deep hole above high-tide level and placed a steel barrel with holes in the bottom inside the excavation. At the time of the highest tide (spring tide), water penetrated through the sand into the barrel, which acted as a stilling well. The water's surface in the stilling well rose and fell with the tide, and its depth could be measured more accurately because there were no waves in the barrel.

The surveys and tide measurements were completed by June 25, 1944. Lieutenant Reid received orders to report to Engineer Headquarters in London to work up their measurements and draw bathymetric charts (maps showing water depths) of Omaha Beach. Another officer, Lieutenant Rechard, who had made similar measurements at Utah Beach, joined him in London.

With measurements over a two-week period, they were able to calculate the first four major tidal constituents. These constituents represent the portions of the tide due to the moon and sun if each were acting independently on a rotating earth, and the effects due to the fact that the moon and sun are not directly above the equator. (The National Oceanic and Atmospheric Administration now lists thirty-seven tidal constituents for San Diego, California, but they have large computers to calculate the data!) Now Lieutenant Reid was able to establish mean sea level at Omaha Beach and could reference water depths along all their survey transects to it. The tidal constituent information also made it possible to forecast future tides along Omaha and Utah Beaches.

Lieutenant Reid and Lieutenant Rechard completed their bathymetry charts by July 7. Their postinvasion maps compared favorably with the preinvasion maps and confirmed that the Keulegan airplane overflight method of determining water depths and beach slopes could be used reliably to plan future amphibious landings.

Bob Reid returned to Normandy on July 8. It had been an interesting two weeks in London, during which he met Charles Bates, John Crowell, and other American, British, and Canadian officers who had participated in the preinvasion planning. It also allowed him some R&R and the opportunity to eat real food instead of their usual ten-in-one field rations (concentrated, dehydrated food packaged to feed ten men three meals per day per package).

~~~

EARLY IN 1941, even before America entered the war, Winston Churchill had the foresight to realize that existing ports would be destroyed during the invasion by Allied bombing or sabotaged by the Germans, rendering them unusable for many months afterward. On the other hand, invading troops would require immediate resupply of food, ammunition, vehicles, and all the other materiel of war in amounts of thousands of tons per day, every day, and could not wait for the ports to be rebuilt. Churchill ordered the planners to design prefabricated temporary harbors that could be towed to the invasion beaches and installed immediately after the landings. These temporary harbors, code-named "Mulberries," would consist of offshore breakwaters, piers for mooring ships, and causeways from the piers to shore. The plan was to use the Mulberry harbors until the French ports could be permanently rebuilt.

After Normandy was selected as the invasion site, the British High Com-
~~~

mand decided that two complete Mulberry harbors would be constructed, "A" for the American sector and "B" for the British/Canadian sector (Figure 10). Each harbor would need to handle ships and equipment equivalent to those serviced by the British port of Dover.[12]

The Mulberries would consist of an integrated complex of offshore breakwaters, floating docks, and floating causeways to connect the docks to the beach. Antiquated surplus ships (code-named corncobs) would be towed to Normandy and sunk offshore to form part of the breakwaters. Specially built concrete caissons (phoenixes) would also be sunk beside the ships to provide additional protection from waves. An outer line of floating steel breakwaters (bombardons) were anchored seaward of the phoenixes. The Mulberries were prefabricated in Britain in a little over six months by a workforce of twenty thousand British laborers. The caissons were designed to be installed in four days.

Intelligence on the beaches was vital to all aspects of the Mulberry artificial harbor project. Special teams of British hydrographers made several sorties to the French coast between November 1943 and January 1944 to make depth soundings and collect bottom samples to assess the bearing capacity of the bottom material. Their first survey took place in late November 1943 at Arromanches, site of Mulberry B. They returned about a month later to make similar measurements at the site of the American harbor, Mulberry A. On June 4, the Mulberry fleet sailed from Britain but had to hold for a day in coastal waters when General Eisenhower postponed the invasion from June 5 to June 6. Installation was well under way by June 9, when Reid and Pritchard landed. Mulberry A was placed off Omaha Beach (American sector), and Mulberry B was off Gold Beach, near the village of Arromanches (British sector). The sixty old ships, called corncobs or block ships, were scuttled about one mile offshore, which along with the concrete caissons, formed the outer breakwaters (see the book jacket). Wave height measurements made inside and outside the breakwaters showed that they reduced the wave heights by 50%.

The floating docks needed to be in water at least twenty-five feet deep so the deepest-draft transport ships could tie up to them even at low tide. This put them approximately 3,500 feet offshore. The docks, made up of sections 480 feet long, totaled about one mile in length. Each section was held in position by an adjustable leg at each corner. The legs were lowered to the bottom and served the same purpose as pilings at a permanent dock. The system was designed

FIGURE 10

Mulberry B in the British sector, looking from sea toward shore at Arromanches. The open waters of the English Channel are toward the lower right. A row of pierheads ran parallel to shore, about 3,500 feet from the beach and formed the wharf to which ships tied up to unload troops, trucks, tanks, and all other types of vehicles. Once unloaded, the vehicles moved across the roadways that stretched back to shore. Specially constructed concrete caissons and block ships (obsolete ships; *beyond far right*) were sunk to form breakwaters (not shown) against the notoriously rough English Channel sea conditions. Mulberry A in the American sector had a similar configuration but was destroyed in the "Big Storm" of June 19–21.

to allow the floating docks and causeways to move up and down as much as twenty feet as the tide rose and fell.

Floating roadways extended from the docks some 3,500 feet to shore. They consisted of "beetles," concrete or steel pontoons that supported sixteen kilometers of "whales," the actual roadways. Each beetle could support the weight of a tank.

Unloading of supplies began at Mulberry A by June 11 and at Mulberry B by June 14, although construction work continued on both facilities. As many as seven transport ships could moor to the docks of each Mulberry at one time.

On June 19, the Big Storm pounded the Mulberries and beaches with thirty-knot winds and waves eight to ten feet high.[13] Rapidly developing meteorological conditions surprised the weather forecasters. Very strong winds from the north and north-northeast sent high waves roaring from the Strait of Dover straight toward the ships unloading at the Mulberries. "We got out a forecast, but not in time to move boats away," John Crowell wrote to the author.[14] This was most unfortunate because major convoys of equipment and men were caught at sea in the storm, and the Normandy beaches were hammered by huge waves.

Commander John Fleming, who had been the staff meteorologist to the naval commander in chief Admiral Bertram Ramsey, Royal Navy, during the D-Day forecasting later wrote Bates that he "experienced sheer misery" at missing the forecast of the Big Storm. He wrote that his only consolation was "that independent analyses of the situation then prevailing have confirmed that my forecast was correct on the evidence available. But the feeling that one had been, however inadvertently, the main contributor to what was almost a major disaster is something not easily put aside."[15]

Mulberry A was destroyed, in large part by bombardons and landing craft that broke free and were slammed into it by waves. Mulberry B was severely damaged, but parts salvaged from Mulberry A were used to repair Mulberry B, which was back in operation within a few days. It was operated by the British until the end of November, when permanent port facilities at Antwerp, Belgium, were repaired and became operational.

Mulberry B was in operation for five months, during which time two million men, half a million vehicles, and four million tons of equipment were unloaded over its floating docks and roadways.[16] In the one week that Mulberry A existed, approximately 309 ships and smaller craft were unloaded, discharging 946 vehicles of every description from jeeps to tanks and 3,500 tons of supplies and equipment, and over 13,400 personnel crossed from ships to shore.[17]

With the destruction of Mulberry A, the Americans were forced to land their troops, equipment, and supplies over the beach from landing craft, surf conditions permitting. This proved both effective and efficient, and at times more was unloaded directly onto Omaha Beach than was unloaded at Mulberry B (see Figure 11). This prompted questions later as to why the Mulberries were required at all.[18]

The supply operations at Normandy supported the massive buildup of personnel and equipment required to retake western Europe from the Germans. As the Allies pushed across France, they required support of every kind, and most of this came over Omaha Beach and Mulberry B until ports at Cherbourg and Antwerp started to become available in October.

Accurate weather and beach forecasts were mandatory for efficient resupply across the beaches. After returning to Normandy from London, Bob Reid rejoined Don Pritchard and their weather detachment. In early August, John Crowell and Charles Bates visited Pritchard and Reid, who by then were posted ten miles inland with the 77th Fighter Division. Reid gave both officers tours of the beaches and the Cherbourg Peninsula and showed them the unloading operations at Mulberry B and at Omaha for which they were forecasting. He even arranged for Lieutenant Bates to make the last DUKW trip of the day for a close-up view of the ship offloading procedure. During his trip to the beach, Bates observed that a growing sea breeze had begun to make unloading difficult and dangerous.

Sea breezes develop, usually in the afternoon, when the land heats up and the air over the land starts to rise. Air above the cooler sea rushes landward to replace the rising air, causing a wind (breeze) that blows from sea toward shore. The wind can be sufficiently strong to cause choppy three- to four-foot waves along beaches.

Lieutenant Bates noticed that crane operators on the ships were having trouble dropping heavily loaded cargo nets into the DUKWs' cargo wells. Unloading of critically needed supplies slowed down and nearly stopped even though there were still five hours of daylight left. Lieutenant Bates located Colonel Moorman at his headquarters in an apple orchard near the battlefront and described the problem to him, explaining that the engineers' unloading rates were

FIGURE 11

American troops, equipment, and supplies unloading from LSTs (landing ship, tank) across Omaha Beach after the "Big Storm" of June 19–21 destroyed Mulberry A. This picture was taken at low tide, and many landing craft can be seen stranded on the beach, waiting for the next high tide before they will be able to leave. Notice the large number of transport ships in the background. Every vehicle and man in the picture (and thousands more) came ashore from those ships via the beached landing craft and hundreds more like them. The balloons were towed behind the ships to discourage low-flying German aircraft from strafing the ships. Courtesy Naval History and Heritage Command, "80-G-46817 Normandy Invasion, June 1944." National Archives ID 80-G-46817.

dropping off drastically because of the unforecasted afternoon sea breeze and similar weather-related problems.

Bates learned that neither the army nor the navy were using detailed wave forecasts. "The missing link appeared to be the lack of correlation between off-loading rate and wave height," he wrote.[19] Bates pointed this out to Colonel Moorman, who authorized him to get the secret unloading data for Omaha

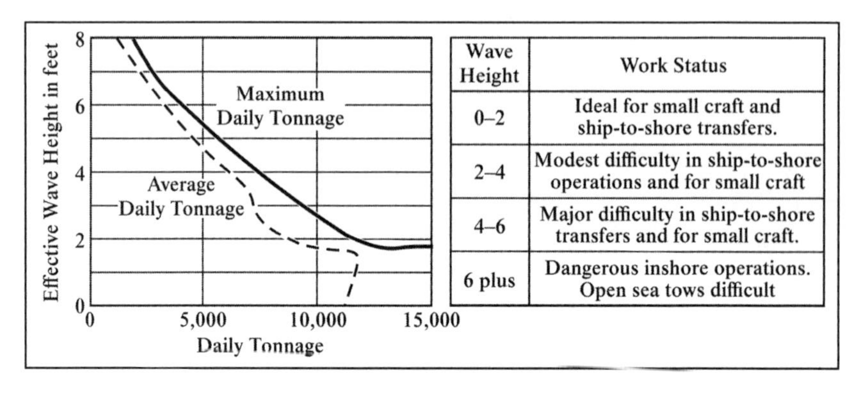

Wave Height	Work Status
0–2	Ideal for small craft and ship-to-shore transfers.
2–4	Modest difficulty in ship-to-shore operations and for small craft
4–6	Major difficulty in ship-to-shore transfers and for small craft.
6 plus	Dangerous inshore operations. Open sea tows difficult

FIGURE 12

Unloading rates versus wave height for Omaha Beach. Adapted from Bates, "Sea, Swell and Surf Forecasting for D-Day and Beyond," 22.

Beach when he got back to London. He plotted unloading rates versus wave height and found that unloading was unaffected by waves less than two feet high. When waves were higher, the unloading rate dropped steadily, and when the waves reached six feet, the unloading rate dropped to about 10% of capacity (see Figure 12).

Colonel Moorman met with the theater's chief engineer, Major General Moore, and they agreed that the engineers should have their own weather detachment near the beach to provide better support to the unloading crews.

On September 7, Lieutenant Pritchard's group of weathermen was officially designated Coastal Weather Detachment YK, with Pritchard assigned as its first commander.[20] His personnel included Bob Reid, another lieutenant, and eight enlisted men. Their mission was marine weather and oceanographic forecasting, including surf and alongshore current forecasts.

At long last, Pritchard's weathermen could leave their tents at the inland airbase and set up shop in the town of Carentan, near the beach and their sole customer, the 2nd Provisional Engineering Brigade. They now had teletype connections with the engineering brigade and weather squadron headquarters and equipment for upper-air soundings. One of their enlisted men, Staff Sergeant Trubeaux, a French-speaking Cajun from Louisiana, made life easier since he could communicate with the local populace.

American forces captured the port city of Cherbourg on June 26, nearly two weeks behind schedule.[21] The Germans had sunk ships in the harbor entrance and had mined it, making the port unusable until the hulks and mines could be removed and the facilities rebuilt. Limited operations began three weeks after the allies captured the port, but Cherbourg did not reach its full capability to handle big cargo carriers until mid-October 1944.

Late in July, the Allied armies began to break through the German defenses and pursued the enemy forces eastward across France. More supplies and personnel were required for the pursuit, but port facilities consisted mainly of the still only partially operational Cherbourg. The Overlord logistics plan anticipated closing out the beach supply operations in late September due to the autumn storms that usually started at this time.[22] The beaches were ordered to continue full-scale operations, however, and proved to be a godsend because of the delayed opening of Cherbourg. In the first seven weeks after D-Day, they were virtually the only source of resupply for the rapidly advancing American armies.

In his two-volume series *Logistical Support of the Armies,* written for the Army Center of Military History, Roland G. Ruppenthal wrote: "Unloading finally came to an end on 13 November at Utah, and on 19 November at Omaha, after 167 days of operation. . . . Fortunately it was possible to operate the beaches considerably longer than originally expected, and their overall record was a spectacular one. In the twenty-four weeks of their operation they received approximately 2,000,000 long tons of cargo, which constituted about 55 percent of the total tonnage brought onto the Continent up to that time. In addition, they had discharged 287,500 vehicles and debarked 1,602,000 men."[23]

The observations by Lieutenant Pritchard's Coastal Weather Detachment YK were an integral part of the twenty-four-hour forecasts of local sea conditions that were prepared for all the beaches and harbors along the coast of France. They contributed directly to the Americans being able to extend the planned life of the over-the-beach unloading operation long enough for the French ports to be repaired.

Describing the work of the Coastal Weather Detachment, Charles Bates wrote, "The detachment proved so skillful at predicting how the environment would control off-loading operations that the engineer beach commander kept operations going full tilt until early November rather than shutting down in late September, as originally planned."[24] Lieutenant Bates's observation of the effects of sea breezes on offloading rates contributed to the forecasting skill of

Detachment YK. The agencies that used its data, engineer units responsible for moving men and equipment from transport ships to shore and on inland, expressed their appreciation for the support provided by the 21st Weather Squadron. Detachment YK continued weather, tide, and surf forecasts from Omaha Beach through November 1944.

No one in the detachment had any objections when they received orders in early December to move to Paris, near the Arc de Triomphe, except perhaps the detachment of enlisted men they left at Normandy to continue the observations. Lieutenant Pritchard and the others in Detachment YK moved their map-plotting and forecasting activities to office space in Paris, complete with French secretaries. They lived in a hotel and ate at an officers' club near the Champs Élysées. Living and working conditions had definitely improved after six months of field life at Normandy. They had Major Seiwell to thank for the new conditions, for he had been able to set up a beach intelligence center in Paris, staffed by forty-two personnel, many of whom were demobilized French experts in coastal hydrography.[25]

The efforts of support personnel like Pritchard and Reid are overshadowed by battlefield stories where the drama and acts of heroism are chronicled, often in gory detail, by journalists embedded with the troops. The fighting troops deserve all the accolades they receive and probably even more, but if not for the soldiers behind the front lines, they would have been stopped dead in their tracks when their vehicles ran out of fuel, their guns ran out of ammunition, or they ran out of food. The soldiers on the front lines owe a debt of gratitude to those who unload the materiel of war and send it forward to them, for they provide the often unheralded support vital to the fighting soldiers.

Lieutenants Reid, Pritchard, Bates, and Crowell received letters of commendation from Major General C. R. Moore, theater chief engineer, endorsed by General Hoyt S. Vandenberg, commanding general of the 9th Air Force, for their contributions to the "rapid advance of our Armies" after the Normandy landings. He singled out Reid and Pritchard for "providing special meteorological forecasts which are proving of great value to the Engineer Service and to the Field Forces." The commendation also noted that one of their "greatest problems was the necessity for convincing using agencies of the need and value of oceanographic information."[26]

5

BEACH TERRAIN

WHEN SOLDIERS AND MARINES ASSAULT a hostile shore, they not only have to dodge enemy bullets; they also have to survive the ocean's nearshore environment. In many World War II amphibious operations, landing beaches were selected primarily on tactical factors. When insufficient consideration was given to beach conditions, it sometimes led to disastrous results.

After receiving orders to conduct an amphibious landing, a commander should reconnoiter the potential beaches available for the assigned operation, considering the mission, the enemy's defenses, and the physical characteristics of the beach. For example, during the war with Mexico, General Winfield Scott sailed along several potential landing sites before selecting Collado Beach for his invasion of Vera Cruz. General Scott chose Collado Beach not only because the Mexican Army had made no defensive preparations there but also because the physical characteristics of the beach were favorable for a landing with his surfboats.

During World War II, aerial reconnaissance was frequently used to learn details of both the enemy's preparations and of the beach itself. Selecting the ideal beach for an amphibious landing can be a tradeoff among many factors: bottom material, bottom gradient (slope), tides, currents, and waves. If a physical inspection or aerial reconnaissance is not possible, the commander should at least conduct a map reconnaissance of the assigned landing beach.

The militarily important oceanographic characteristics of a beach are surf conditions, bottom gradient seaward of the beach, the general configuration of the bottom offshore from the beach, protection from winds and waves, longshore currents, offshore bars or reefs, tidal fluctuations, and the bearing capacity of the beach material. Surf conditions and bottom slope are perhaps the most important military characteristics of a beach. Heavy surf capable of interfering with amphibious operations can be caused by severe local storms or by swell that has propagated into the area from distant storms. As waves approach shore, their forward motion slows, their length decreases, and their height increases. In planning an amphibious operation during seasons when severe storms are probable anywhere near a commander's assigned area of operations, the commander should keep himself advised of all large storms and their possible effects on his operation. Even though the storm may not be predicted to actually enter his geographic area, its effects, especially wave action, can be felt many miles distant from the storm center. Studies of wave data recorded at Barbados in 1958 indicated the sudden appearance of swell generated by a hurricane in the North Atlantic. Another study in 1963 tracked storm-generated swell from the Southern Hemisphere almost halfway round the world.[1]

Bottom slope and surf conditions are intimately interrelated. A gently sloping bottom causes the waves to start breaking a long way from shore and consequently to dissipate their energy over a wide area. Waves approaching a steep shore race closer to the beach before breaking and hence concentrate their energy in a narrow, highly turbulent zone close to shore.

Most landing craft used during World War II were essentially shallow-draft boats and could proceed up a beach only as far as the water was sufficiently deep for them to remain afloat. Once they touched bottom, they could go no farther. A very gently sloping bottom can cause landing craft to ground far from shore and thus expose troops to enemy gunfire while wading ashore with no cover or concealment. A steeply sloping bottom allows landing craft to get close to shore before grounding, minimizing the distance the troops must wade without cover or concealment. On the other hand, a steep bottom gradient can produce surging or plunging breakers that are more violent than the spilling breakers found on a gently sloping beach. Examples of both favorable and unfavorable surf and bottom slope conditions can be found in historical accounts of US Army and Marine amphibious operations.

Waves created by winds blowing at an angle to a beach can cause currents to flow parallel to the beach. These currents, known as longshore currents, can reach speeds of two to three knots and can sweep heavily laden men off their feet and push landing craft thousands of yards from their intended landing sites. Longshore currents were particularly troublesome for the duplex drive tanks at Normandy on D-Day. These tanks were not very seaworthy and broadside currents of two knots or more easily pushed some of them thousands of yards off course as they slowly worked their way shoreward through several miles of surf. Many of the duplex drive tanks foundered before they reached the beach, but most of those that did make shore found themselves on the wrong beach, among soldiers of the wrong units. The tank commanders had been briefed on their support missions when they got ashore, but when they landed in the wrong place, they were at a loss as to how to improvise and support the unit where they were. Infantry units were deprived of the artillery support they expected and needed from the tanks, greatly contributing to the heavy casualties suffered by many of the units on the Normandy beaches.

The bottom material on the ideal beach would be firm sand. Soft mud, large rocks, coral heads, or holes in the bottom all slow the wading troops and their vehicles and assist the defenders. Extensive soft mudflats at Inchon, Korea, and a coral reef at Tarawa Atoll in the Pacific are examples of extreme bottom conditions that aided and hindered landings during the Korean War and World War II, respectively.

Bottom Gradient and Breaking Waves

The bottom gradient of a beach is the slope of the bottom—how quickly (or slowly) the water gets deeper as you go seaward from the shore. The bottom can slope very gently away from the shoreline, such as along the west coast of Florida, or it can be quite steep. A steep slope can make an attractive landing site for an amphibious assault since landing craft can get close to shore before dropping their ramps and discharging the troops, thus minimizing the distance the soldiers must wade through water under enemy gunfire. On a beach with a gentle bottom gradient, the landing craft grind to a stop a long distance from shore, forcing the troops to wade much farther than they would if the beach gradient were steep. Every extra step the wading soldiers must take is obviously a serious problem if the enemy is dug in on the shore and firing at them.

Waves approaching shore react differently to steep or gentle bottom gradients. As discussed in chapter 2, the circular orbital motions of waves in deep water extend downward into the water column and gradually decrease in diameter with depth below the surface. At a depth of half the wave length, the wave motions are so small they are virtually undetectable.

As the deepwater waves travel shoreward into shallower water, the orbital motions reach the bottom and the waves begin to slow down. When the waves feel bottom, they increase in height and begin to break. On a gently sloping beach, waves begin to feel bottom far from shore. The crest grows in height as the wave advances shoreward and becomes concave on both sides. The thin crest is unstable and begins to fall apart, spilling down the wave front and forming foam that is pushed along in front of the wave as it continues shoreward. Spilling waves frequently re-form and continue in to the beach, and there may be several lines of foam, representing the number of spilling waves that have broken on the beach. Spilling waves usually pose no serious hazard to an amphibious landing craft (see Figure 13).

A gentle bottom gradient causes the landing craft to hit bottom a considerable distance from shore, and consequently the marines and soldiers have to wade farther through enemy gunfire to reach the beach. A gentle bottom gradient can present other problems, as was the situation faced by the Australian 7th Amphibious Force when it landed at Balikpapan, Borneo, July 1, 1945. The beach gradient was so gentle that their fire support ships could get no closer than five miles from the beach. Landing without their fire support contributed to a very difficult amphibious landing.[2]

Over a steeply sloping bottom, incoming waves rush toward the beach with little change until they rather abruptly feel bottom. They slow down and increase in height, and the wave crest curls over, resulting in a concave shape to the front of the wave. The "curled-over" water is literally hanging out in space and crashes into the trough of the previous wave. Waves of this type are called plunging breakers. Expert surfers love these long overhanging waves, which they call a "tube," as they precariously surf along under the overhang.

Plunging breakers can be extremely hazardous to an amphibious landing, however. When a landing craft, or any other type of boat, gets into the surf zone and attempts to run to the beach, it risks broaching. A boat broaches when it gets into a wave environment and cannot keep pace with the wave.

The rear face of a wave is more rounded than the forward face, and inexperienced boaters or landing-craft coxswains have run up waves without realizing

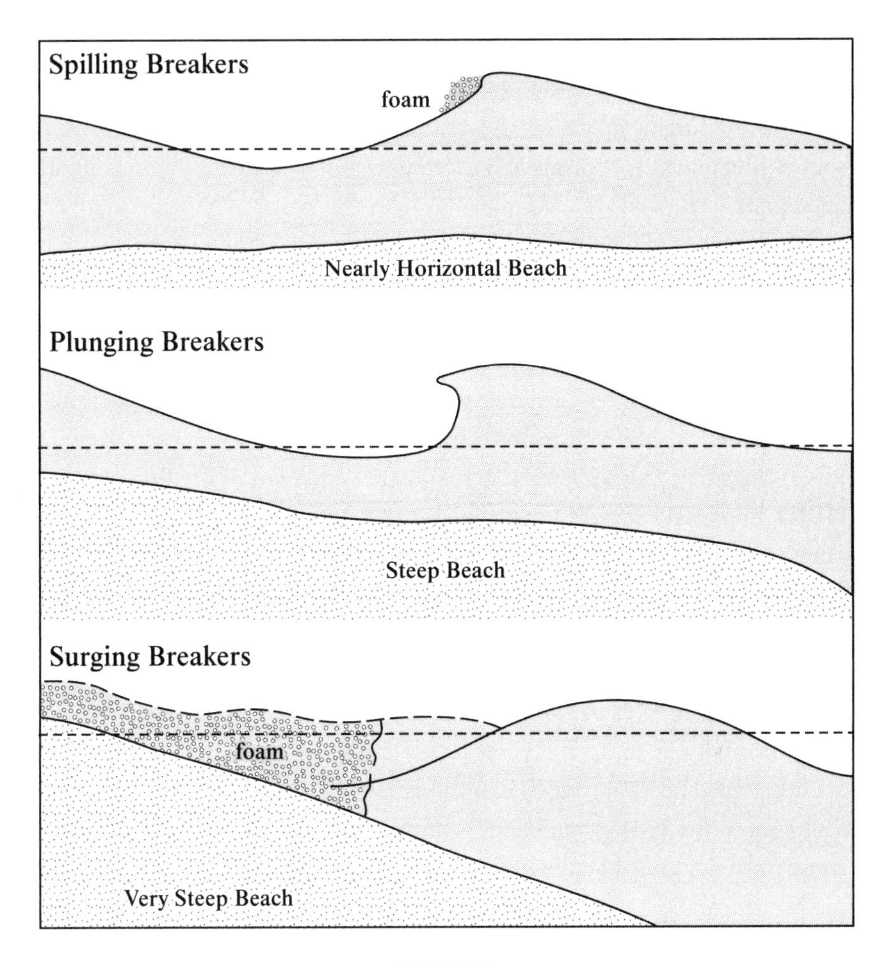

FIGURE 13
Spilling, plunging, and surging breakers.

the danger they were getting into. If they run up the wave until they near the crest, the wave's orbital motions are moving in the same direction as the wave and the boat may suddenly be carried over the crest. At this point, the bow of the boat is no longer in the water and falls into the wave trough ahead. The boat's stern is swept forward and the boat will usually broach and capsize, all hands likely lost.

A beach with a very steep bottom gradient can produce surging breakers. In this type of breaker, a plunging breaker starts to form, but due to the steepness of the beach gradient, the bottom of the wave rushes forward ahead of the forming wave crest. This stops the plunging, and the wave collapses in a mass of water and foam surging up the beach face. Surging breakers are not as common as spilling and plunging breakers and have rarely posed problems for amphibious operations.

During the landings at La Paz, Subic Bay, Philippines, the surf at all beaches was extremely high. Crews of several LCVPs lost control of their boats, which then broached. A navy control boat sank unexpectedly. Self-propelled barges and even two LSTs were swept ashore by the heavy surf. By the fifth day, seventeen of forty-three landing craft were unserviceable. With surf conditions so rough, lighters had to use more protected landing sites along the Dagupon River. The surf conditions were so bad that completion of the first phase of the landing was delayed from S plus six to S plus ten days.[3] (S-Day was another designation for the start of an operation—similar to the familiar D-Day at Normandy.)

During the liberation of Borneo, beach conditions were generally much better, although choppy seas made towing pontoons difficult. Also, loading of pontoon barges from LSTs was difficult and dangerous in the heavy seas.[4]

Difficulties due to steep bottom slope were experienced in unloading LSTs at Catmon Hill Beach during the consolidation phase of the operation at Leyte, Philippines. The LSTs grounded ten to fifteen yards from shore, requiring the construction of ramps before offloading could commence.[5] At White Beach in the Leyte operation, the slope was too gentle for LSTs to land properly, so they were rerouted to Cataisan Point, where the steeper slope made dry landings possible.[6]

The general configuration of the bottom seaward of a beach can have significant effects on the surf at the beach. If the bottom slope is uniform along the beach, an approaching wave slows uniformly along its entire length as it runs into shallower water. If, on the other hand, the bottom has ridges or troughs, the portion of the wave that feels bottom first slows while the remainder of the wave continues shoreward at its original speed. This will cause the wave crest to bend toward the shallower area and concentrate its energy there with a concomitant decrease of energy in the deeper area. This process is known as wave refraction (see Figure 14). Thus a feature a commander might consider in select-

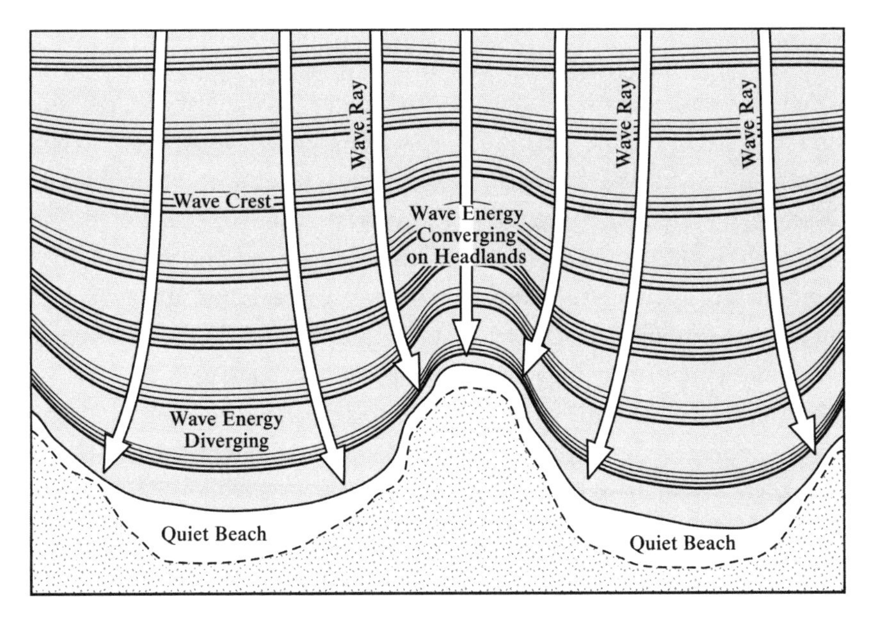

FIGURE 14

This figure illustrates the process of wave refraction, which is caused by uneven bottom topography. As waves that are parallel to one another offshore (at the top of the figure) continue shoreward, the portions of the waves that enter shallower water begin to slow down while the remainder of the wave front continues shoreward at its original speed. This has the effect of bending the wave front toward the shallow water. Note how the waves seem to "wrap" around the headland protruding outward in the center of the figure. Another way to think of this is that at the top of the figure wave energy and wave heights are evenly distributed between the wave rays (white arrows). As the waves approach shore and experience varying water depths, they "bend" toward shallower water and concentrate their energy there, with higher waves. In areas where the wave rays diverge, the wave energy is spread over a greater area and wave heights decrease, leading to the quiet beaches shown.

ing a beach is bottom topography, a steep slope directly in front of the beach and shallow ridges extending seaward on either side of the deep area being preferable. Waves approaching such a beach would be bent toward the ridges with decreased wave heights in the center. This would require extensive, expert beach reconnaissance that would not be generally available.

Longshore Currents

When the wind blows at an angle to the shoreline, friction between the wind and the water moves the water in the general direction of the wind. This is both toward and parallel to the beach. The wind also creates waves that strike the beach at an angle. Both winds and waves cause longshore currents that flow parallel to the beach (see Figure 15). These longshore currents flow within the surf zone and can reach speeds of two to three knots. Currents this strong can make wading ashore difficult and can push landing craft and other vehicles hundreds of yards (or more) away from their intended landing locations.

After the US 6th Army secured the Philippine island of Leyte in October 1944, supplies were moved ashore over floating pontoons and docks. During this Leyte consolidation phase, longshore currents caused great difficulty in holding floating docks in place at Red Beach.[7]

As the US and Philippine armies were retreating from Manila in early 1942, the Japanese Army tried to use longshore currents to locate the entrenched and camouflaged Americans and Filipinos along the east coast of the Bataan Peninsula. They waited until night, then launched bamboo floats carrying batteries and flashlight bulbs into the longshore current, which they knew flowed past the beach where the Americans and Filipinos were dug in. The officers in charge of the entrenched soldiers ordered the men to hold their fire. The Japanese tried

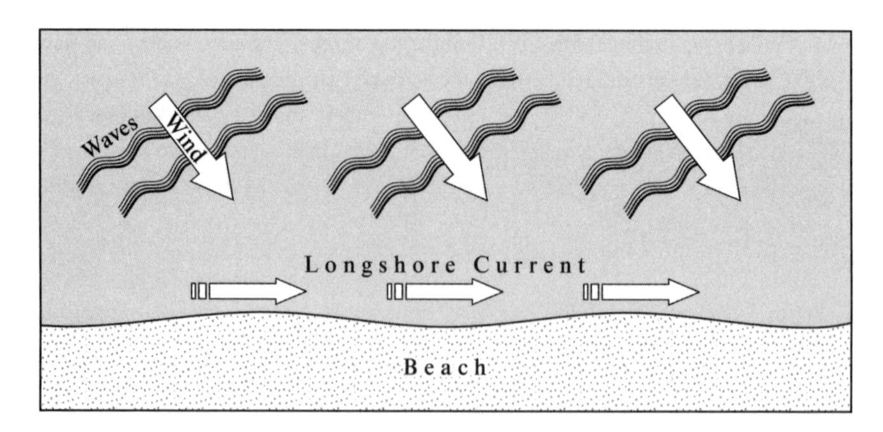

FIGURE 15

Wind blowing at an angle onto a beach creates waves and longshore currents. The currents occur within the surf zone and flow parallel to the shoreline.

again over the next few nights, but when the Filipinos and Americans still did not fire at their lights, the Japanese towed a barge with machine guns on it past the beach and fired in the general direction of the soldiers. This time they could not resist firing back, but they paid a heavy price. The Japanese plotted the locations of the muzzle flashes ashore and called in artillery fire on the American and Filipino positions.[8]

Tide

Tide period (time between tides) and range (height of tide) are important factors to be considered for amphibious operations. Tides are caused by the gravitational attraction of the moon and sun (and other celestial bodies) on the waters of the ocean, as discussed in detail in chapter 4. The range of the tide can vary considerably even at a given point. The highest tide ranges, spring tides, occur twice each month when the gravitational forces of the sun and moon are either in conjunction or opposition. The lowest tide ranges (also twice monthly), neap tides, occur after the first and last quarters of the moon. Tides can be predicted with a high degree of accuracy for any area, given previous tide data from that area on which to base the forecasts. Accurate tide predictions are vital to a commander planning amphibious operations in an area that has an appreciable tidal range.

Tidal conditions were of great concern to General MacArthur and his staff during planning for the invasion at Inchon, Korea. Inchon had "every conceivable and natural handicap" to an amphibious operation.[9] Embayments, such as the harbor at Inchon, can amplify the ocean tide and create conditions inside the bay much more severe than those outside. The average spring tide range at Inchon is twenty-three feet and reaches as much as thirty-three feet. Tidal currents in the channel approached five knots, and the channel was a maze of rocks, shoals, reefs, and islands. The "beach" at the end of the channel was a twelve-foot-high seawall, and at ebb tide the entire area was a vast mudflat extending as far as three miles to sea. Only on September 15 and October 11, 1950, would the spring tides be high enough for LSTs to negotiate the channel and land at the beaches and then for a period of only three hours either side of high tide.[10] These were indeed formidable obstacles to be overcome, and General MacArthur relied heavily on precise timing and the element of surprise to land successfully at such an unlikely location. Surprise was achieved, however, and the US X Corps (which consisted of the US 1st Marine Division and the US Army 7th Infantry Division) landed at Inchon under light enemy resistance. The landing was a spectacular success, against great natural odds, and was accomplished only through thorough planning and knowledge of the oceanographic conditions, especially tides, at Inchon.

The effects of other beach characteristics can be enhanced (or lessened) due to the stage of the tide at the time the operation is conducted. A good example is a bar that might be crossed at high tide but could be exposed at low tide. Another example is strong tide-created currents in an inlet, which could make navigation difficult (as at Inchon). Large waves can be created when an outgoing tidal current is opposed by heavy surf from winds blowing into the inlet. When the tide changes and the current flows into the inlet (in the same direction as waves and winds), the waves can diminish. A commander should be aware of these factors and the timing according to tide stage and take them into consideration during planning for amphibious operations.

Knowledge of nearshore and tidal currents was put to a different use as part of the deception plan for the invasion of Sicily. The body of a man dressed as a Royal Marines major with a courier's briefcase chained to the body was taken to the coastal waters off the southern coast of Spain near the town of Huelva by a British submarine. Papers inside the briefcase were intended to deceive the Nazis into believing that any landings on Sicily were merely deceptive actions to screen the main invasion, which would take place in Greece and Sardinia.

Planners of the operation needed information on tides and currents along the southwest Spanish coast in order to select a release point from which they could be assured the body would reach shore and fall into German hands. They turned to the Royal Navy's Hydrographic Department and were informed that tidal currents ran "mainly up and down the coast" and the prevailing winds at that time of the year were from the southwest. They also had the latest information on tidal currents from the superintendent of tides, which they included in their operation orders for the mission.[11] With this information, a release point was selected, the "major's" body was loaded on the submarine, and it sailed for the selected location off the Spanish coast. The body was released at night and was discovered the next day by local Spanish citizens, who delivered the briefcase to German intelligence sources. The deception was effective, and the incident is credited with aiding the invasion of Sicily.[12]

Bottom Material

A beach may be composed of almost any naturally occurring material on earth, such as mud, sand, cobbles (or cobblestones), coral heads, coral sand, and rocks of all sizes from gravel to boulders. In the first stages of the planning process, the commander must determine the bearing capacity of the bottom material on the assault beach, especially when vehicles will be landed. Beach reconnaissance is necessary but usually difficult because the enemy wants to deny knowledge of his beach.

If the bottom is packed sand, the planners probably will not need to give it a second thought. Men and vehicles can cross such a beach with little difficulty. A packed-sand beach is also a fine recreational beach. Automobile races were even staged on the hard-packed fine sand of Daytona Beach, Florida, from 1905 until 1958. In contrast, the coarse sand on the beaches between Pensacola and Panama City, Florida, is loose, and any cars attempting to drive on the beach usually become stuck. This happened to military vehicles on some of the Pacific island beaches during World War II. Beaches selected for the landing on Morotai Island (near New Guinea) were selected based on interpretations of black-and-white aerial photographs. The photographs were interpreted as showing "white coral sand veneer on a gently sloping, hard coral rock shelf." Actual conditions were quite different. The reef was covered with soft, thick sediment and the coral underneath was full of large holes. The men had great difficulty getting through the soft sediment and "heavy equipment sank almost out of sight" as soon as it left the ramps of the LSTs.[13]

Natural Beach Obstacles

The degree to which a beach is protected by headlands or offshore reefs and bars can decrease surf conditions significantly. If the larger ships can sail into waters protected by a surrounding land mass, men and materiel can be loaded into landing craft and landed on the beaches under relatively low surf conditions. Rougher seas in open water then would be of less concern to the army commander.

Offshore reefs and bars also can offer a degree of protection to the landing beach if there is a way to go around them. Crossing a reef or bar can be hazardous not only due to the danger of grounding but also because the shallow water over the bar causes waves to break there. Consequently an offshore reef or bar can be either a hazard or a benefit. In the Pacific operations during World War II, most accounts of reefs and bars indicate that they were impediments to the operations.

At Okinawa, a reef extended along the entire beach, parallel to the shoreline. It was impossible to cross the reef with naval craft, so all personnel and equipment had to be transferred to landing vehicles, tracked (LVTs) and DUKWs, which caused delays in the landing.[14] At La Paz, a landing beach was changed because an offshore bar made landings difficult.[15] The planned beachhead at Leyte was narrowed from two thousand to one thousand yards because of the shallow water in the center of the landing area and the enemy fire. Intelligence forecasts prepared by the Corps of Engineers indicated that a sand bar fronted the beach two hundred to three hundred feet offshore. Evidently the forecast was disregarded because the landing took place there anyway, and three LSTs grounded on the bar and suffered heavy structural damage and casualties.[16] The landings on Palawan also encountered difficulties with an offshore reef. Pilots of the first four waves of boats became confused by the reef and landed as much as three hundred yards from their designated landing points. Only the fifth wave landed properly. Boats carrying artillery grounded on the reef and later had to be towed ashore by other boats. Fortunately for the invading task force there was no enemy opposition and the sea was calm. Heavy surf on the reef undoubtedly would have caused many casualties and the loss of much equipment. A hydrographic survey conducted after the landing showed this beach to be very unsatisfactory for landings.[17]

At Licata Beach and Red Beach during the invasion of Sicily, the 7th Army encountered problems due to beach slopes, offshore bars and surf. High seas caused by gale-force winds delayed the naval task force and disorganized it. Ships were in wrong positions, which caused longer runs to the beach in landing craft and also caused some troops to land on the wrong beaches. The heavy surf added considerably to the difficulties encountered by the army forces during the landing. Since the beaches in this region of Sicily were generally sandy with occasional rock outcrops, they appeared ideal for landings. However, they were not. The beach slope was too gentle for many of the larger craft to get as close to the shoreline as they needed to be. Shifting sand bars were present along much of the beach and prevented many of the larger, heavily laden craft from reaching shore. The army forces that went ashore at Yellow and Blue Beaches fared better due to more favorable beach conditions and less resistance.[18]

Beach Reconnaissance and Intelligence

The need for good beach intelligence prior to conducting any amphibious operation is obvious. There are numerous reports of beach reconnaissance parties being sent out before and during landing operations to obtain beach information. Hydrographic studies made prior to the capture of Lae, Philippines, indicated that deep water extended close inshore and there were no coral formations on a beach about eighteen miles east of Lae, which lead to the selection of the beach for the landing.[19] At Corregidor, crash boats were sent under fire to survey the feasibility of landing LSMs (landing ships medium) on the beach.[20] Sixteen days prior to the landings on New Britain Island, a patrol of amphibious scouts reconnoitered the Cape Gloucester area, and landing beaches were selected on the basis of this patrol.[21] At Morotai Island, hydrographic surveys were conducted to select sites for wharves and docks.[22] During operations on Leyte, a hydrographic survey unit ran over two hundred miles of sounding lines to determine a deep channel through Cancabato Bay into Tacloban Harbor. A channel twenty-one feet deep at low water was found and marked by this unit.[23] A US Navy officer, Lieutenant Eugene F. Clark, was sent ashore to inspect beach conditions prior to the landings at Inchon.[24] His exploits are discussed in detail in chapter 7.

During World War II, the Beach Erosion Board (BEB), predecessor of the Coastal Engineering Research Center, US Army Corps of Engineers, played an important role in supplying beach intelligence to the US armed forces, particularly the army. The BEB was established in 1930 and, until the outbreak of World War II, contributed significantly to the study of beach processes and to the protection and preservation of the nation's beaches. The BEB completed its first intelligence report, *Landing Area Report: Cherbourg to Dunkirk,* in July 1942 and thereafter found itself involved in the war effort.[25]

The BEB's efforts were concentrated in two areas: the preparation of beach landing reports and wave research. In preparing the landing reports, information on beach characteristics was collected and transferred to maps of the area under consideration. Then a report was written to describe the area and included any available photographs and charts.[26] The beach and the approaches to it, from about thirty feet of water depth to the hinterlands and exits from the beach, were detailed on the maps.[27] The BEB, under Brigadier General John J. Kingman, senior member of the board, and Dr. Martin Mason, the senior civilian, prepared beach landing reports on many important areas, including Normandy, the Philippines and other Pacific islands, Sicily, Italy, and North Africa.

Wave research proceeded almost continually in the BEB's wave tanks. Dr. Garbis Keulegan was in charge of this activity, which investigated oceanographic problems such as remotely determining water depths over offshore bars, the development of movable breakwaters, water depth determination by studying changes in wave lengths as recorded by aerial photography, and sediment transport. The research on movable breakwaters showed them to be effective according to Dr. Keulegan,[28] and several types were used during the Normandy invasion.[29]

The BEB also prepared strategic planning reports, which were used principally for strategic planning rather than actual operations.[30] They were prepared from strips of aerial photographs to show major terrain features and were used primarily to indicate areas that were or were not feasible landing sites.

The BEB trained civilian personnel in the techniques of gathering beach information to be analyzed for beach intelligence. After training at the BEB, these specialists were sent to the theaters of operations, mostly in the Pacific, to train army personnel in beach intelligence work. In addition to training troops, they planned a number of surveys and laid out specifications for the data collection, which was done by military personnel.[31] Since these advisors were civilians, they were not supposed to take part in the operations, although they did go in after the beachhead was established to verify the accuracy of the intelligence reports.[32] There is one account of an advisor going ashore only three hours after the start of the operation to determine report accuracy.[33]

The value to the war effort of the beach intelligence work and other contributions by the BEB is verified by the fact that several of the key individuals received Civilian Service Awards from the War Department at the conclusion of the war.[34] Another indication of the value of their studies is that according to Dr. Mason, "Everyone seemed to want studies done. We had more work than we could do."[35]

6

NEAP TIDE AT BLOODY TARAWA

FOR HUNDREDS OF YEARS, TARAWA, a coral atoll in the western Pacific Ocean, was the idyllic tropical island paradise portrayed in movies. Then in 1941 the Japanese Army arrived and turned it into an armed fortress. The Japanese built an airfield on Betio Island, part of the Tarawa Atoll, to support their World War II naval operations. The Americans wanted the field for the same purpose, so in November 1943 the US Navy and Marines arrived to take it away from the Japanese. The ensuing battle turned Tarawa into what author John Wukovits called "one square mile of hell."[1] When it was over three days later, 5,680 Americans and Japanese had died ghastly deaths, and the island was a rubble-covered graveyard—all for a pile of coral sand less than half the size of New York City's Central Park.

Although the Tarawa landings preceded those at Normandy, planning for the invasion of Europe was well under way by the time fighting ended at Tarawa on November 23, 1943. Some of the "lessons learned" at Tarawa made their way back to London and undoubtedly contributed to the operations plan for Operation Overlord, but many were specific to the island-hopping plan for war in the Pacific Theater. There also were delays in communications between the Pacific and European Theaters of Operation due to the great distances separating them and the strict security restrictions. Suffice it to say that General Eisenhower's planning staff in London did not have immediate access to the lessons learned at Tarawa, and Lieutenants Bates, Crowell, and Cauthery had no feedback on forecasting oceanographic conditions.

Tarawa is part of the Gilbert Islands, a group of sixteen coral atolls and islands named for Captain Thomas Gilbert, the British mariner who discovered it on June 20, 1788.[2] The Gilberts straddle the equator about four hundred nautical miles west of the International Date Line, which puts them about halfway between Hawaii and Australia. The three most populous and important Gilbert atolls are Makin, Tarawa, and Abemama. The first settlers in the Gilberts are thought to have been Micronesians who arrived from southeast Asia some three thousand years ago. Life on Tarawa was a languid existence of subsistence fishing and coconut gathering, interrupted by occasional interisland disputes.

Life for the Gilbert Islanders was unhurried but required hard work. The islands are small, and few rise more than twelve feet above sea level. The soil on the islands is mostly coral sand that supports very little plant life other than coconut palms and pandanus trees. The natives used every bit of both. Coconuts and fish were their main foods, supplemented by the fruit of the pandanus. The tree trunks were used to construct homes and other buildings. The pandanus leaves were woven into mats and hats and were used to thatch the roofs and sides of buildings; the roots were thought to have medicinal properties. The coconut husks were used as thread to bind the thatches and for fishing lines. The only other edible plant was a taro-like plant that took much work to grow. The sea provided a reliable source of food, but fresh water was usually in short supply. A somewhat serendipitous result of the lack of natural resources was that few foreigners who arrived on the islands found much reason to stay and left the natives to their languorous existence.

Life in the Gilberts changed rapidly between 1830 and 1890 as increasing numbers of Europeans arrived. The earliest visitors were whalers who stopped for water, fresh food, and the sailors' eternal need—women. As contact with whalers increased, a more stable relationship developed between the Gilbertese and Europeans. The first missionaries arrived in 1850, and life changed again, mostly for the better as the missionaries established schools and churches.[3] A permanent European community had grown up by the 1860s and established a profitable trade selling coconut oil, copra, dried turtle shells, and additional island-produced goods to other Europeans and Australians. Under the influence of the British missionaries, the Gilbertese adopted Christianity.

A dark chapter in the story of European visitors was the "labor traders," actually slavers who kidnapped the islanders and took them away to be sold as slave labor.

Robert Louis Stevenson and his wife made three cruises through the South Seas, including the Gilberts, between 1888 and 1890. He was so smitten by the Gilbert Islands that they lived on Abemama Atoll for several months. Stevenson wrote, "I remember it best on moonless nights. The air was like a bath of milk. Countless shining stars were overhead, the lagoon paved with them." He also said the islands had "a superb ocean climate, days of blinding sun and bracing wind, nights of a heavenly brightness."

In the decade prior to the Stevensons' arrival, missionaries had persuaded the native women to give up the custom of going completely unclothed until they married. It may be that the most significant problem left facing the missionaries at that time was trying to convince them that they needed to cover their top halves also.[4]

In 1892, Britain declared the Gilbert Islands a British protectorate, then a colony in 1915. The islands remained a British colony until granted self-rule by the United Kingdom in 1971 and gained full independence in 1979 as the Republic of Kiribati.

~~~

IN THE LATE 1930S, few Gilbertese had any reason to pay attention to, or even knew about events taking place thousands of miles away in Japan and China. These countries had fought each other and Russia for centuries without their conflicts affecting the Gilbert Islanders. That would soon change and embroil not only the Gilbert Islands but most of the rest of the world.

Toward the end of the nineteenth century, Japan invaded China several times and extended its economic and political influence over the Chinese. Generalissimo Chiang Kai-shek organized the National Revolutionary Army to resist the Japanese, and by 1928 had achieved a tenuous unification of China. At this same time, the Chinese Communist Party grew stronger. Chiang Kai-shek and the Communists cooperated against the Japanese for a while, but their ideological differences soon drove them apart.

Meanwhile, the Japanese military grew stronger and wrested control of the country away from the civilian government. Japan continued to increase its political and economic dominance of China until war broke out in 1937 between the Empire of Japan and the Republic of China. Japan extended its influence
~~~

into French Indochina (Vietnam, Cambodia, Laos), where it began to build military bases in the late thirties.

By 1940, the western governments—the United States, Britain, New Zealand, and Australia—had become alarmed over Japan's aggressive expansion of influence. To curtail this expansion, they agreed to stop selling iron ore, steel, and oil to Japan. The embargo was a serious blow, which Japan viewed as aggression. Unable to support its far-flung conquests with the embargo on raw materials, Japan began preparations in May 1941 to secure these commodities through war. They planned to attack Pearl Harbor, then seize the Philippines, Hong Kong, Malaya, and most of the western Pacific island groups. Part of Japan's strategy was to isolate Australia and New Zealand from the other Allied powers.

The attack on Pearl Harbor on December 7, 1941, initiated the plan and was designed to destroy or neutralize the US Navy. In rapid succession after Pearl Harbor, Japan attacked and conquered Hong Kong, Guam, Wake Island, the Philippines, Thailand, Singapore, and Malaya. By late 1942, Japan occupied most of the western Pacific island groups that had any military or economic significance. However, Japan miscalculated the reaction of the United States to this aggression, and its attack on Pearl Harbor brought the United States into World War II.

The Japanese began building a defensive perimeter of strategically placed military bases around these conquered lands so the western navies could not interdict their shipments of oil and raw materials to Japan. Australia and New Zealand would soon be isolated.

The Gilbert Islands, and Tarawa in particular, were strategically located to be part of Japan's defensive ring. An airfield on Tarawa would enable Japanese land-based airplanes to support army and naval operations over a vast expanse of the western Pacific. Three days after the Pearl Harbor attack, Japanese Army units arrived at Tarawa to begin constructing an airfield, defensive fortifications, and barracks. They chose Betio Island, part of the Tarawa Atoll, for the airfield and their main base in the Gilberts.

Betio resembles a great crested bird, its beak facing north, its crest to the southwest, and its tail pointing southeast. The bird's back is toward the open Pacific Ocean to the south, while its breast and belly border the lagoon. A long pier completes the bird's outline as its legs extending into the lagoon (see Map 3).

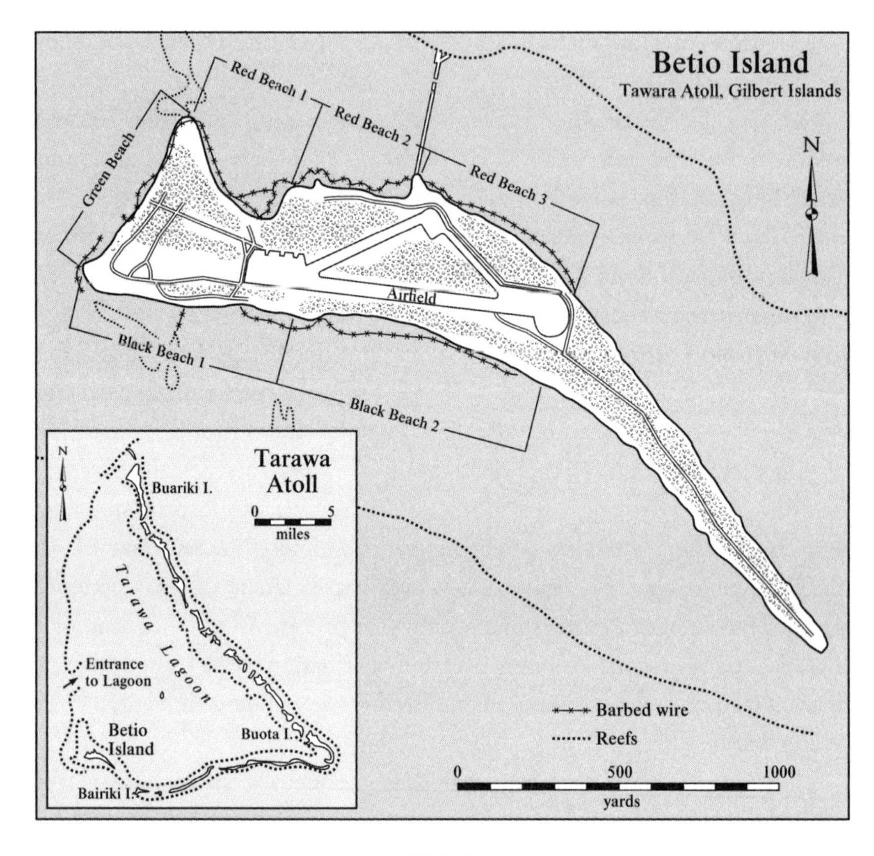

MAP 3

Tarawa Atoll (inset) and Betio Island. The dotted lines surrounding Tarawa represent the coral reef that is on both sides of all islands and encloses the lagoon. The only entrance to the lagoon is a narrow passage on the western side. The map of Betio Island shows the airstrip, the pier, barbed-wire obstructions in the water, and the landing beaches where marines stormed ashore on November 20, 1943. Adapted from University of Texas Map Collection (inset); HyperWar: A Hypertext History of the Second World War (Betio Island).

Japanese Army construction units, assisted by conscripted Korean laborers, built a runway that ran 4,750 feet from east to west, from the bird's crest to the base of its tail. Then they started work on an extensive array of fortifications and barriers. They built a seawall of coconut logs three to five feet high behind all the likely invasion beaches of Betio (see Figure 16). These were to serve as barriers to foot soldiers and tanks. They placed gigantic concrete tetrahedrons (similar in shape to children's jacks) in the water, cleverly spaced to divert landing craft into point-blank fire from shore guns hidden in reinforced concrete bunkers.

The most effective barrier was the coral reef itself. It encircled the island and was less than three feet deep at certain tide stages. The Japanese thought American landing craft could not cross it.

There were two eight-inch coast defense guns on each end of the island that could play havoc with any ships or landing craft that approached from the sea or the lagoon. They were well camouflaged and protected by barricades of concrete and coconut logs. Dozens of smaller guns were placed to fire at anything that approached the seawalls and the airfield, and antiaircraft guns bristled over the island. Approaches to the airfield from the beaches were blocked by antitank trenches. Mines had been placed in the water along the southern and western beaches, and mining of the approaches to the lagoon beaches was under way in November 1943.

The Japanese strung barbed wire fences in the water around the island and along the beaches to impede attackers wading ashore from landing craft. All of these barriers were covered by machine guns to mow down anyone who tried to cross them.

Pillboxes, ammunition dumps, gun emplacements, and command and control shelters were constructed of heavily reinforced concrete up to five feet thick. Layers of coconut logs were crisscrossed over the concrete, and then coral sand was piled many feet deep on top of the logs. Coconut logs proved to be very resilient defensive materials that absorbed much of the shock of an artillery shot rather than splintering and breaking up like most other types of logs. Five hundred fortified positions were tied together with trenches to permit soldiers to move among them.

Rear Admiral Keiji Shibasaki, commander of the Japanese forces on Betio, bragged: "A million men cannot take Tarawa in a hundred years."[5] He had more than 4,800 soldiers on Tarawa to back up his boast. About three thousand of these troops were from the elite and superbly trained Yokosuka 6th Special Naval Landing Force and the Sasebo 7th Special Naval Landing Force. These were equivalent to US Marine Corps units.

FIGURE 16

Aerial photograph of Betio Island in the foreground, looking east, taken September 18, 1943, by an American photo-reconnaissance aircraft two months before the assault. The reef around each individual island is clearly visible as indicated by the light-colored water, with deep (darker) water beyond the reef's edge. The lagoon is to the left, and the open Pacific Ocean is to the right. The Japanese airstrip was under construction when this photograph was made and is visible down the center of Betio Island. Courtesy Naval History and Heritage Command, "80-G-83771 Tarawa Atoll, Gilbert Islands." National Archives ID 80-G-83771.

THE AMERICAN JOINT CHIEFS OF STAFF (JCS) and their British counterparts held several conferences in early 1943 to decide the best way to carry the war to the Japanese. In May 1943 the Combined Chiefs of Staff approved operations to island-hop across the central Pacific to position themselves to invade the Japanese homeland.

On July 20, 1943, the American JCS approved Operation Galvanic, the first step in the island hopping, with amphibious landings in the Ellice Islands and Gilbert Islands and at Nauru Island. A month later, Makin Atoll in the Gilberts was substituted for Nauru.

Planning for Galvanic jumped into high gear because the JCS wanted the landings in the Gilberts to start on November 15. Bases in the Gilbert Islands would serve as springboards to the Marshall Islands, and from there on to other island groups approaching Japan. Any delays in taking the Gilberts would give the Japanese more time to prepare defenses in the Marshalls and islands farther west.

The 2nd Marine Division, stationed in New Zealand, was designated to attack Tarawa, with marine Major General Julian C. Smith in command. The army's 27th Division, training in Hawaii, was assigned to take Makin, under army Major General Ralph C. Smith. Tarawa and Makin were to be attacked simultaneously.

Marine amphibious doctrine called for the first wave of an amphibious landing to hit the beach on a rising tide, several hours after daybreak, to allow time for accurate naval and aerial bombardments of the landing area before the troops arrived. The landing should not take place late in the day because there would be insufficient daylight left to establish a beachhead before dark.

As soon as the planners looked at the details of the tides and ocean conditions at Tarawa, they realized they needed to adjust their doctrine. Tide tables for Tarawa in November 1943 showed high tides occurred several hours before daybreak and late in the afternoon, which was unacceptable. The invasion could be delayed until late December when tides would align with their doctrine, but waiting would delay all the subsequent landings in the island-hopping scheme. The planners decided to delay the assault until November 20 and go in on a neap tide, when the water levels were expected to remain nearly constant for twenty-four to thirty-six hours. Thus the ocean dictated November 20 as D-Day at Tarawa.

Training and rehearsals were stepped up, and large numbers of vehicles were readied. Space was limited, so training was spread over five amphibious bases in Hawaii and several beaches in New Zealand. Special training with the landing craft was conducted in far-off Fiji, which had beach characteristics similar to those in the Gilbert Islands. Rehearsals also took place at Efate Island, about halfway between New Zealand and the Gilberts.

With these far-flung preparation and training sites, there was insufficient time to stage all the troops and equipment in one place for shipment to Tarawa. Therefore, the operation plan specified that the US Navy would transport the marines and their equipment and landing vehicles to Tarawa. Then, on the morning of the attack, the navy would unload everything into the landing craft to take the marines the last three miles from the ships to the beaches of Betio Island. This is the same procedure General Scott used at Vera Cruz.

Two types of landing craft were available for the operation: the LVT (see Figure 17) and the LCVP. Using LVTs to land troops on a heavily defended shore was a new concept. Before Operation Galvanic they had only been used to take equipment and supplies ashore after the assault had secured the beach. However, the Marine Corps had conducted extensive tests that showed that LVTs could be used by troops to storm a beach.

The LVT, or amphibian tractor, also known as an amtrac, was propelled by tracks, similar to a tank, and could travel over land as well as through water. The planners knew that the marines would have to cross the offshore reef, but they were not sure how deep the water would be. Water depth would not be a problem for the amtracs because they could drive up to the reef, and if the water was too shallow, climb over it, drop into deeper water on the other side, and continue to the beach. When they reached the beach, they could drive inland and unload on dry land. The amtrac became the preferred landing vehicle for Tarawa.

Major General Julian Smith was able to get seventy-five operable amtracs from his own inventory and scrounged up fifty more in San Diego. There was not time to mobilize these vehicles and their crews from San Diego to New Zealand to rehearse with the invasion troops, so the landing craft were shipped directly to Samoa, where their drivers and maintenance crews joined them. Ad-

FIGURE 17

Landing vehicle, tracked (LVTs) were an open amphibious tractor, sometimes called amtracs, and were used to carry equipment and men from ships to beaches. Its tracks propelled it on water or land. Early versions like the one pictured had very little armor or firepower. The gunner, with two .50-caliber machine guns, was completely exposed, and the driver was protected by nothing more than a Plexiglas windshield. Here, assault troops jump from a LVT(2) amphibious tractor as it hits the beach on Morotai, September 15, 1944. Courtesy History and Heritage Command, "80-G-257947 Morotai Operation, September 1944." National Archives ID 80-G-257947.

ditional 3/8-inch armor plate was hastily installed on these LVTs. The marines scheduled to go ashore in these amtracs would see their landing vehicles and drivers for the first time on the morning of the invasion.

But even 125 LVTs were not enough. General Smith needed one hundred to land the first three waves of marines, and he needed some in reserve. He knew that many of these LVTs would be disabled or destroyed while landing the first three waves. Therefore he would have to use LCVPs to land the last two waves.

FIGURE 18

The landing craft, vehicle, personnel (LCVP), also known as the Higgins boat, was the workhorse landing craft of World War II and used in virtually all amphibious landings. Unlike the LVT, it was a true boat and could not climb over the reefs on Tarawa. Pictured here, LCVPs are landing troops during invasion practice at Slapton Sands, England, in 1944. Courtesy Naval History and Heritage Command, "80-G-252326 North European invasion rehearsals, 1943–1944." National Archives ID 80-G-252326.

LCVPs (see Figure 18) were used at Normandy and most amphibious assaults in World War II and also at Inchon, Korea. They were true boats, driven by an underwater propeller, and required water at least four feet deep. They could carry more cargo and men than the amtracs, but being boats, they could go no farther when they hit bottom. If the water over the reef was not at least four feet deep on D-Day morning, the LCVPs would stop when they reached the outer edge of the reef and would have to unload the marines three hundred to seven hundred yards from shore.

Because of the water depth limitations of the LCVPs, the question of tides

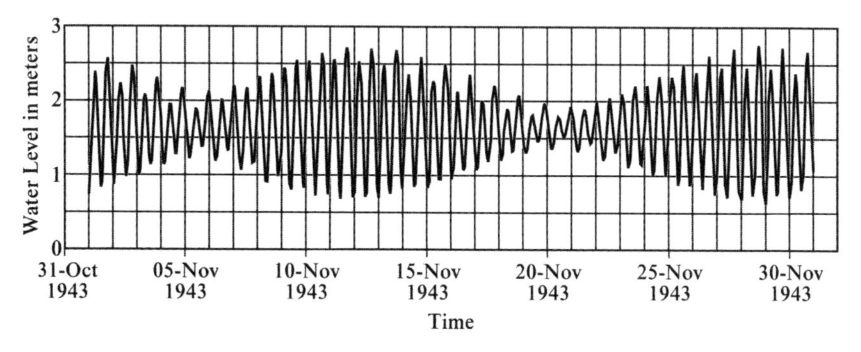

FIGURE 19

Tides at Tarawa for November 1943. There are two nearly equal tides on each day. Spring tides (November 11 and 29) and neap tides (November 5 and 20) are apparent. The water levels (in meters) are referenced to lowest low water, zero on the figure, which is the lowest water level the ocean ever reaches at that location.

and water depth over the reef quickly became one of General Julian Smith's top concerns. The best charts available in 1943 had been made by an American hydrographer a hundred years earlier. They were incomplete and unreliable. The Allies' main reconnaissance resources were aerial photographs and photographs made through submarine periscopes. Neither provided any information on water depths at the reef.

Figure 19 shows the tide range (in meters) at Tarawa for the month of November 1943. High tides reach about 2.7 meters higher than the reference level (zero).

The intelligence staff at Pearl Harbor began searching for mariners, ex–Gilbert Islanders, or anyone else who had any knowledge of the tides and waters around Betio. They assembled a group of thirteen or fourteen people, mostly Australian and New Zealand naval officers, and took them to staff headquarters at Pearl Harbor to develop tide tables for Tarawa. This group of Gilbert Island resident experts predicted that high neap tide on November 20 would produce water depths of five feet over the reef. If this was true, the LCVPs would have enough water under their keels to get them over the reef.[6]

The Gilbert Island experts repeatedly warned of a phenomenon they called "dodge tides," a term unfamiliar to American or British hydrographers or oceanographers. It was said that dodge tides could make neap tide predictions difficult and unreliable. The planners were so concerned about dodge tides that they even mentioned the phenomenon in the operation plan: "During neap tides, a 'dodging' tide has frequently been observed when the water ebbs and flows several times in 24 hours."[7]

The term seems to have assumed a certain mystique in the literature about Tarawa, but it is simply a local South Australian term for a neap tide with minimal change in water depth (tide range) for twenty-four to thirty-six hours.[8] The mystique revolves around what causes dodge tides to ebb and flow several times in a tidal cycle and how they can be forecasted.

The most likely force causing irregular variations in water depths is wind. Even a moderate wind will cause water to flow and in shallow water particularly, depending on wind direction relative to shore, can alter water depth. Wind directed shoreward will cause the water to pile up on the shore (deepen), or wind directed offshore will draw the water down (shallowing). This effect would be accentuated in a small enclosed body of water such as a lagoon. It also would be more noticeable at neap tide when the water level is not expected to change much for twenty-four hours. It is likely that the mysterious dodge tide is nothing more than the ocean's response to the wind changing speed or direction during a neap tide. On the lagoon side of Betio Island, an east or south wind would cause the water to become shallower; a north or west wind would deepen it.

Major F. L. G. Holland, a New Zealand Army officer who had spent fifteen years on Tarawa, was among the Gilbert Island experts in Hawaii. He also warned of dodging tides and at first agreed with the prediction of five-foot water depth over the reef. He recanted when he learned the marines planned to land at neap tide.[9] Major Holland said his prediction was for high spring tide, and he never dreamed that anyone would land on a neap tide. He said there would be less than three feet of water over the reef, even at high neap tide. The American planners decided that keeping the invasion momentum going was worth accepting the risk of landing on a neap tide and disregarded Major Holland's concerns. They continued with their plans to land on a neap tide.[10]

~~~

THE WEATHER AT TARAWA on the morning of November 20, 1943, was excellent. Skies were partly cloudy with cloud bases at two thousand feet. Surface winds were from east-southeast at twelve knots (13.8 mph), which would tend
~~~

to blow water out of the lagoon, lowering the water depth over the reef at Betio. Waves along the ocean beaches were two to four feet high and were lower in the lagoon, where the main invasion would take place.

The attack began at 0613 hours with approximately fifteen minutes of aerial bombardment of the island. Then naval gunfire began from four battleships, five cruisers, and nine destroyers and lasted for seventy-five minutes. A brief pause in the gunfire followed as the ships were repositioned. The next phase of the naval bombardment began at 0745 hours and lasted for about one hour. During this phase of the preparatory fires, the first three waves of marines in amtracs began their forty-five-minute runs to the beaches. At 0850 the naval vessels ceased firing and the aircraft returned to strafe the beaches with machine-gun fire until 0900. All firing ceased then and the first amtracs hit the beaches at 0910 hours.[11]

Major Holland was right about the water depth. The moon was at three-quarter phase, and a neap tide occurred as was forecasted. "In no place was there more than three feet of water over the reef, and in spots the depth was a matter of inches," Navy historians wrote.[12] The first waves of amtracs reached the outer edge of the reef but simply climbed over the coral as they were supposed to do and continued their runs to the beach.

Once the bombs and naval shelling stropped, the stunned Japanese quickly recovered and manned their gun positions along the beaches. Warrant Officer Kiyoshi Ota, one of the few Japanese survivors, occupied a machine-gun position west of the pier. He had confidence in the men around him and their preparations, but he had not seen amtracs before, and when they climbed over the reef, one of their principal barriers, he knew there was little hope of a Japanese victory.[13]

Murderous enemy machine-gun fire greeted the marines after they crossed the reef. The US Navy commanders had been confident that few Japanese would survive the preattack bombardments, but the naval and aircraft fire were not nearly as effective as everyone thought they would be. Although General Smith had requested three days of naval and aerial bombardment, he received less than three hours. Rear Admiral Richmond Kelly Turner, the naval commander, wanted to surprise the Japanese as much as possible and reasoned that an extended bombardment of the Betio beaches would give the Japanese time to mount strong air attacks on the transport ships before the landing craft could put the marines ashore. Therefore he allowed only three hours of preparatory naval and aerial bombardment.[14] The amtracs came under heavy machine-gun and antiboat fire from the time they reached the reef all the way to shore. From two hundred yards offshore inward, the gunfire became increasingly accurate and damaging. Overhead shell bursts rained shrapnel into the boats. Mortar shells obliterated entire amtracs and their occupants. Boats were disabled and began to drift, making them easier targets for the gunners ashore. Landing craft that were supposed to stay together went off course, spreading members of the same unit over several hundred yards of beach. Officers were separated from their men. All semblance of unit organization and cohesion was lost. At least twenty-five amtracs were sunk or disabled in the initial assault.

The first amtracs reached shore at Red Beach 1 at 0910 hours, near the bird's beak, with the 3rd Battalion, 2nd Marines. They were hit hard by machine-gun fire, and most of the vehicles were damaged. Those marines lucky enough to be in amtracs that reached the beach clambered out of their vehicles and raced for the log seawall, which provided some protection from the Japanese gunfire. Many didn't make it, though. In the first two hours, these two companies lost half their men. Another company had to start wading when their boat was hit and could go no farther. The company lost 35% of its men before they reached the relative safety of the seawall. Five of six officers in one company that landed near the pier were killed or wounded at the beach.

When the landing craft reached shore, the marines often had only a few yards of sand to cross between the water and the seawall, but the Japanese had the beaches well covered with machine-gun and mortar fire. It was extremely difficult to run across even this narrow stretch of sand without being wounded or killed. Bomb craters on the beach afforded about the only protection available until they reached the seawall. Those who made it to the seawall crouched behind it and were very reluctant to leave this temporary safe haven. As soon as they raised their heads above the wall, bullets screamed by from every direction. The Japanese built the seawalls as barriers to the invaders, which they were, but they also provided about the only cover the marines had on Betio's beaches.

Many of the amtracs were disabled in the water, and most of those that reached shore were stopped by the seawalls. A very few found openings in the walls and raced inland as far as the airstrip, where their marines dismounted.

This is what they had planned for most of the vehicles to do, but those few that made it through found themselves all alone.

Some units were able to gain small footholds just beyond the seawall but usually only fifty to seventy-five yards inland. Attempts to move forward were repulsed by strong Japanese machine-gun fire. These units were essentially trapped.

As soon as the first three amtrac waves were on their way, the final two waves of marines loaded in the LCVPs and headed for shore. At approximately 1000 hours they reached the reef but could go no farther when their LCVPs grounded in the shallow water. All the marines could do was unload at the reef's outer edge and wade four hundred to five hundred yards to shore through heavy rifle and machine-gun fire. The marines ducked down as low as they could, with only their helmets above the water. Maj. Michael P. Ryan, assistant battalion commander on Red Beach 1, remarked that with only their helmets visible above the water the marines reminded him of tiny turtles battling against the current.[15]

In an attempt to explain the failure of the LCVPs to cross the reef, a navy historian wrote, "The tide suddenly and dramatically failed. There was inadequate water at the outer edge of the reef."[16] This is incorrect. The tide did not fail. The tide-producing forces worked as reliably that day as they had for eons. It was the planners who failed to predict the shallow water and failed to somehow get more amtracs.

Some of the men, heavily laden with packs and equipment, stepped into deep potholes in the reef and drowned. Amtracs that unloaded at the beach and were still functional returned to the reef and shuttled as many marines as possible from the LCVPs to shore. Confusion reigned as units became separated or intermingled, officers lost contact with their units, and communication was lost. Radios soaked in seawater or riddled by bullets didn't work very well.

One landing team commander, Major John F. Schoettel, was on his way toward the reef in an LCVP when his commander, Colonel David Shoup, ordered him to land his reserve units. "We have nothing left to land," Major Schoettel replied.[17] Most of the units already ashore had not only nothing left to land but very little left of what they had already landed.

The safest route to the beach, but only relatively so, was along the western side of the long pier. The LCVPs began to run to the pier to unload. Then the marines either were picked up and shuttled shoreward by amtracs or they waded alongside the pier, which was too high above the water for them to climb onto it. Japanese snipers on shore had a shooting gallery with human targets as the marines ducked from piling to piling along the pier.

It seemed that there were snipers in every palm tree. Japanese soldiers had climbed the trees and tied themselves to the tree trunks among the palm fronds. Here they had good vantage points and were well hidden and fired at the marines until they were finally spotted and shot.

As bad as the situation was for the marines, it was tougher for the Japanese. When the marines finally climbed over the seawall and attacked, they began to destroy the Japanese pillboxes. From the outside, these reinforced concrete bunkers looked like large mounds of sand. Some contained as many as a hundred soldiers. Entrances to the bunkers were covered by machine-gun fire to prevent attackers from getting in the door. Interior walls were constructed so that any enemy who did get inside found himself wandering in a maze with no clear shot at the Japanese occupants, who then quickly dispatched the intruder.

The pillboxes had one vulnerability, though: air vents on their tops. The marines quickly devised a way to exploit this weakness. A team of marines concentrated heavy machine-gun fire on the defenders outside the structure. As soon as the Japanese guns were silenced, several marines with a flame thrower or cans of gasoline climbed to the top of the pillbox. They poured the gasoline in the air vents or shot a flamethrower into it and then threw several grenades into the vent. That usually killed or immobilized the Japanese occupants, but to make sure, a team of riflemen rushed in the door with rifles blazing and fixed bayonets to finish off any survivors. The flamethrower was one of the most awful and feared weapons of close combat, and it was said that one could suck the air right out of a man's lungs.

As darkness fell that first day, the marines dug in and set up defensive perimeters wherever they happened to be. Some held a line inland about seven hundred feet from the tip of the bird's beak. Others had penetrated nearly to the airstrip in the vicinity of the pier and held about twelve hundred feet of sand on either side of its base. These were the only consolidated positions held by the Americans. The rest were dug in in small pockets, mostly in bomb craters along the beaches.

After that first day's battle, the Americans fully expected the Japanese to counterattack after dark, but nothing happened. They learned later that Admiral Shibasaki had been killed that afternoon and the Japanese communications had been destroyed by the bombing, leaving them leaderless and with no way to organize a counterattack. The marines passed a fearful but relatively quiet night. Some managed to get a little sleep and awoke the next morning thankful to still be alive.

So ended day one of the fight for Bloody Tarawa.

~~~

THE KILLING CONTINUED for three more days, with the American marines gradually gaining the upper hand. The marines began to sense despair spreading among the Japanese defenders and a shift in momentum to the American's side. By midafternoon on the second day, Lieutenant Colonel Presley M. Rixey, commander of an artillery detachment, wrote, "I thought up until 1300 today it was touch and go, then I knew we would win."[18] At 1600 hours, Colonel David M. Shoup, General Julian Smith's operations officer, sent this terse situation report to General Smith: "Casualties: many. Percentage dead: unknown. Combat efficiency: We are winning."[19] The Japanese launched four desperate counterattacks during the night of November 22 but were repulsed in brutal hand-to-hand fighting that left hundreds of soldiers dead on both sides.

On November 23, the marines drove the remaining Japanese to the tip of the bird's tail. Seventeen were captured, but the remainder chose to die fighting or committed suicide. At about 1300 hours, an American carrier-based airplane landed at the airfield—the reason for all the carnage in the first place. "Is it over?" The pilot asked the cheering marines who crowded around his plane.[20]

When the stars and stripes were hoisted on Betio later that afternoon, 4,690 Japanese soldiers and 990 American marines were dead. Of the original 125 amtracs, only 35 were operational at the end of the siege (see Figure 20).

The sickly sweet stench of human death hung over Betio. Now began the task of burying the bloated, sunbaked bodies and pieces of bodies that littered the island. Many lay where they had fallen in the sand or had floated in the ocean for three days. Some were too mutilated even to identify which nation sent them to Tarawa. "Tarawa was the most heavily defended atoll that would ever be invaded by Allied forces in the Pacific," Vice Admiral George Dyer wrote.[21]

~~~

FIGURE 20
Day two of fighting on Red Beach 2 at Tarawa. The amtrac in the background was disabled as it tried to climb over the coconut log seawall, probably by the Japanese machine gun position almost directly in front of it. Note the decapitated palm trees. The arrows point to bodies of dead US Marines.

THE AMERICANS TOOK the airfield at Tarawa, but the price was high—too high, many said. Why were there so many marine casualties? Delays in loading the amtracs pushed back the landing time twice and caused mass confusion among the amtracs offshore. The prelanding naval gunfire and air support periods were too short and were not timed to the amtracs' arrival on the beach.

The intelligence staff in Hawaii predicted five feet water depth at the reef's edge on November 20, as published in the operation plan. The US Coast and Geodetic Survey and the British Hydrographic Office have since published data that say it was four feet. This is a remarkably accurate forecast by the Gilbert Island resident experts, considering they had only scanty hard information with which to work and were relying mainly on memory.[22]

The operation plan makes it clear that even at the highest command levels it was anticipated that LCVPs would ground at the reef's edge and marines would have to wade or be shuttled ashore. The shuttling slowed the landings and the all-important factor of momentum was lost. The momentum could have been maintained if more amtracs had been available. Landing on the neap tide may not have been optimal but was a calculated risk recognized and accepted by the JCS. The decision may have been correct, but the calculation of water depth over the reef was faulty. Most military historians and analysts have concluded that more lives were saved by keeping the momentum of the island-hopping scheme going than would have been saved at Tarawa by waiting for a spring tide.

Landing on a high spring tide might have presented the marines with another serious problem. The ocean would have been lapping at the base of the seawall or even partway up it. With the water deep enough for the LCVPs to run up to the seawalls and lower their ramps, Japanese machine gunners would have waited for the ramps to drop, then would have filled the boats and marines with bullets. Very few Americans would have stepped ashore. Amtracs would have been more appropriate even on a spring tide. Major General Holland Smith, commander, V Amphibious Corps, said, "Without the amphibian tractor, it is believed that the landing at Tarawa would have failed."[23] He just needed more amtracs; and more air strikes and naval shelling would have helped.

One American historian described Betio Island as the most heavily defended island the US forces ever attacked. It also was the first all-out frontal assault that US forces made on such a target. The US Marines were successful in taking Tarawa from the Japanese, and although the price was high in terms of American lives lost, many valuable lessons were learned that were applied in future amphibious landings made by the US Marines and Army.

The lessons learned at Tarawa were particularly useful to operations in the Pacific Theater of Operations, where the "island-hopping" scheme consisted of attacking targets similar to Betio Island. Many of the lessons were tactical in nature: the navy needed a dedicated amphibious command ship because radio communications were interrupted when the big guns on the command ship fired; unloading of supplies needed to be controlled by the tactical commander ashore rather than the amphibious task force commander; better communications were needed between tank and infantry personnel; more backpack flamethrowers and more LVTs were needed; there was a need for underwater swimmers to report reef, beach, and surf conditions (this was the start of the underwater demolition teams); radios needed to be waterproofed; better coordination between naval and aerial preparatory gun fire with the first waves of landing craft was needed; and aerial and naval preparatory fire was not as effective against strongly fortified targets as planners thought it would be.[24]

7

HIGH SPRING TIDE AT INCHON

THE KOREAN WAR FOLLOWED CLOSE on the heels of World War II. The same amphibious warfare doctrine and equipment were used, with most of the landing and shipping craft having seen service in Europe and the Pacific theatres of the previous war. Most if not all of the officers and commanders who served General MacArthur in Korea had seen extensive service in World War II, as had, of course, General MacArthur himself. General Walton H. "Johnnie" Walker was MacArthur's 8th Army Commander and had served under General George Patton in Europe during World War II. MacArthur's deputy, Major General Edward M. "Ned" Almond, had served with distinction in both World War I and World War II. The marine and navy commanders at Inchon had amphibious landing experience from World War II landings, mostly in the Pacific. Hundreds of other army, marine, air force, and navy officers had served in World War II. The situation was different for the enlisted men, though. When World War II ended, a majority of them got out of the service as fast as they could and had no interest in reenlisting or joining the reserves. Therefore, a large percentage of the enlisted ranks were new draftees with only minimal basic training and no combat experience.

On June 25, 1950, approximately 135,000 North Korean soldiers swarmed across the 38th parallel, the border between North and South Korea (see Map 4).[1] The furious attack by the North Korean People's Army (NKPA), or Inmingun, was unexpected, and the ninety-eight thousand ill-equipped, ill-trained, outnumbered soldiers of the Republic of Korea (ROK) and their five hundred US advisors were pushed southward by the relentless attackers from the north. By

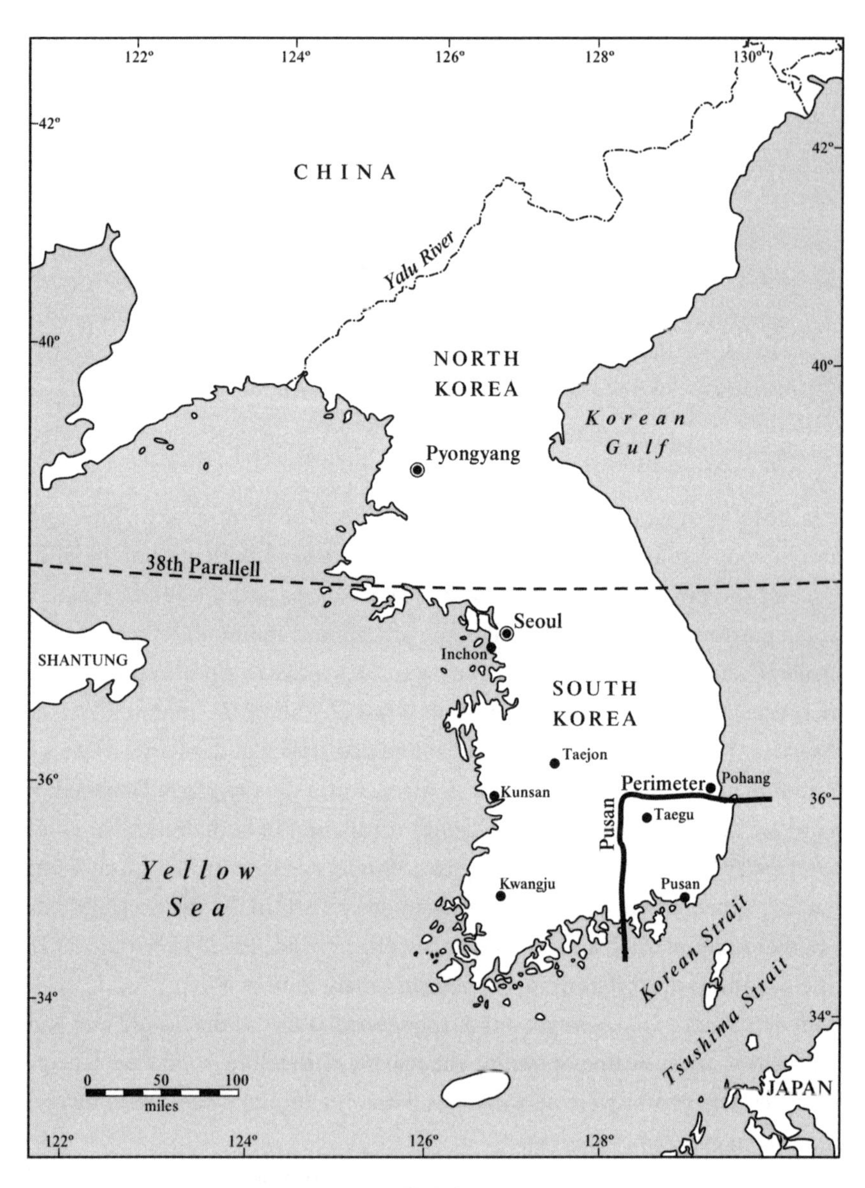

MAP 4

Korea, September 1950. The 38th parallel was the dividing line between Communist North Korea and the Republic of Korea (South Korea). By August 4, all UN forces in Korea had been driven into a small area around Pusan, within the Pusan Perimeter. Operation Chromite landed US Marines at Inchon. Then they drove across the peninsula, recaptured Seoul, and continued to the east coast.

mid-August, the ROK Army, the American and other UN advisors, and four to six million South Korean refugees had retreated until they were crammed into an area roughly 140 miles long by 40 miles wide. This rectangle, which included the port city of Pusan, was dubbed the Pusan Perimeter. The situation was so dire that the American Joint Chiefs of Staff started contingency planning to evacuate the American, UN, and ROK forces by sea in what many feared would be a repeat of the Dunkirk debacle of World War II.

General Douglas MacArthur, who had commanded the Allied forces in Japan since the end of World War II, was appointed commander of UN military forces in South Korea. He had not forgotten his ignoble flight from the Bataan Peninsula in the Philippines during World War II when he abandoned thousands of American and Filipino troops who were captured and imprisoned by the Japanese. He did not want to repeat this humiliating chapter in American military history and sought a way to free his 8th Army, trapped at Pusan by the Inmingun, and win a quick victory in the process.

Although the NKPA was a formidable fighting force, its supply lines were long and unprotected. They stretched the entire length of the Korean Peninsula from China to Pusan, making the army vulnerable to air attack, so the NKPA moved its supplies at night to reduce this vulnerability. General MacArthur proposed making a massive amphibious landing along the Korean west coast at Inchon, a port city about twenty-five miles west of the capital city of Seoul. From here he would drive inland to sever the NKPA's supply lines. An army deprived of steady resupply could not survive long, and he envisioned a quick end to the war.

General MacArthur's staff warned him that Inchon had many natural hazards, but he was undeterred in his determination to attack there. He had conducted many successful amphibious landings during World War II, and he was confident that the marines could succeed and would not let him down. He had learned what was required of an amphibious attack and would assemble the requisite men and equipment. World War II had ended only five years earlier, landing craft were still available, and he would prepare them in time to storm the waterfront at Inchon. General MacArthur was convinced that the merits of an amphibious assault at Inchon far outweighed the many drawbacks his staff, and others, foresaw.

The JCS and President Truman agreed with General MacArthur's strategy until they learned that he planned to land at Inchon, which is about 180 miles north of Pusan. They recommended landing at Kunsan, another west coast seaport only about fifty miles from Pusan that had fewer operational and logistical challenges. The JCS was so concerned about MacArthur's plans that they sent General J. Lawton Collins and Admiral Forrest Sherman to Tokyo to confer directly with MacArthur about the details of his plans and to talk him out of the idea. They arrived in Tokyo on August 21.

General MacArthur argued that the high tides, vast mudflats, and other geographical obstacles at Inchon gave the North Koreans a false sense of security. Since they considered Inchon such an unlikely location for an amphibious landing, it was lightly defended, and this made it the right place to attack. The NKPA's main rail and road supply lines converge on Seoul, which is only about twenty-five miles east of Inchon, so the American X Corps would have only a short distance to advance to recapture the capital city and sever the NKPA supply lines to all of their forces in the south. As they headed east from Inchon, they planned to capture Kimpo airfield just west of Seoul. Restoring control of the South Korean capital to the country's president would be a huge morale boost to all South Koreans, especially the ROK Army.

General MacArthur convened a full-scale briefing at 5:30 p.m. on August 23, 1950, in his office in the Dai Ichi Insurance Building in Tokyo. The general planned to lay out his plans for the Inchon invasion for the JCS visitors and put to rest any opposition to his plan. The meeting was attended by more than a dozen senior officers, including General Collins and Admiral Sherman and the admirals and generals whose troops would have to carry out the invasion.

MacArthur began the meeting with an overview of the general situation. Then, one by one, the officers, most of whom were trying to change MacArthur's mind, rose to describe the myriad difficulties of invading at Inchon and recommended landing elsewhere. Some were concerned that Inchon might be another Tarawa, where the World War II invasion of the Pacific atoll cost the marines at least 990 men in three days of savage fighting.[2] General Collins later wrote that the mood of the meeting was "pessimistic."[3] Summing up his opinion of the Inchon invasion, Rear Admiral James H. Doyle, the very experienced amphibious commander for the operation, turned to General MacArthur and said, "General, I have not been asked nor have I volunteered my opinion about this landing. If I were asked, however, the best I can say is that Inchon is not impossible."[4]

The JCS representatives, General Collins and Admiral Sherman, and the navy and marine commanders tried their best to persuade General MacArthur to change his mind and invade at Kunsan or some other more favorable location instead. Lieutenant Commander Arlie G. Capps, gunfire support officer for Rear Admiral Doyle, has been widely quoted as saying, "We drew up a list of every conceivable natural and geographic handicap and Inchon had 'em all."[5]

General MacArthur listened quietly to the opinions and advice of his senior officers, then launched into an eloquent forty-five-minute monologue as to why Inchon was the right place to attack the NKPA. He even admitted that it was a five thousand to one gamble, but he remained adamant that the very objections his staff raised were why the plan would work. "The history of war proves that nine out of ten times an army has been destroyed because its supply lines have been cut off. We shall land at Inchon, and I shall crush them," MacArthur said.[6]

Collins and Sherman returned to Washington and briefed the JCS and President Truman. On August 28 they sent General MacArthur a rather ambiguous cable that read, "We concur in making preparations and executing a turning movement by amphibious forces on the west coast of Korea, either at Inchon in the event that enemy forces in the vicinity of Inchon prove ineffective or at a favorable beach south of Inchon if one can be located."[7] This provided a small crack in the JCS opposition to Inchon, and apparently that was all the leeway MacArthur needed. On August 30 he issued an operation order to launch the amphibious landing at Inchon, with D-Day set for September 15. He did not send a copy of the order to the JCS until September 8. By then, D-Day was one week away.

~~~

ONE OF THE BASIC PROBLEMS in planning the amphibious invasion at Inchon was the lack of reliable information about the seaport. Navy Captain Edward Pearce was in charge of the Geographic Branch of General MacArthur's staff in Tokyo and was responsible for gathering intelligence on tides, terrain, and facilities at Inchon. The branch staff had been working around the clock to gather all the intelligence they could find on Inchon, but they still lacked very basic and vital information. They had both American and Japanese tide tables for Inchon,
~~~

but they did not agree. Which, if either, should be used? Inchon was fronted by vast mudflats at low tide, but no one knew if they would support vehicle traffic, or even foot soldiers. How high were the seawalls around the city? Were there camouflaged gun emplacements that could not be seen from the air? Where were exits from the beach into the city? What were the best routes for trucks and troops out of Inchon toward Seoul? Had the North Koreans mined the channels leading to Inchon? Major General Charles Willoughby, MacArthur's intelligence chief had expressed serious reservations about the intelligence reports the Geographic Branch had issued. Aerial reconnaissance had revealed a lot, but these questions remained and Captain Pearce knew they could only be answered by firsthand observation. Fortunately, he had an officer on his staff, navy Lieutenant Eugene F. Clark, who was ideally suited for this assignment.

Since the war began, Lieutenant Clark had worked for Captain Pearce in the Geographic Branch, and he and his team of experts had gathered every scrap of information they could find about tides, terrain, and landing facilities along both coasts of Korea. They scoured old Japanese sources, aerial photographs, and information from World War II. Clark realized there were serious gaps in their knowledge, but the information they needed simply did not exist.

Captain Pearce called Lieutenant Clark into his office on August 26, 1950, and told him that General MacArthur was planning an amphibious landing at Inchon on September 15, and the general considered it essential that they obtain more timely and accurate information on everything in and around the place—at once. Clark's World War II amphibious experience made him the logical choice to get it since he knew what information was needed and how it would be used.

The invasion was news to Clark, but the veteran of Okinawa and several other World War II amphibious operations immediately recognized the significance of the plan. It would sever the NKPA's lines of communication and supply with their armies in the south. He also recognized that on-site observation was the only way to get the information, and he relished the opportunity to leave the office in Tokyo and get back in the field. Clark knew he would need help with the language, and in dealing with the South Koreans he would need to recruit from the local population to help him gather the information General MacArthur needed.[8] The next morning, August 27, he flew to Taegu, within the Pusan Perimeter, to recruit two South Korean naval officers, Lieutenant Youn

Juong and Colonel Ke In-Ju. Clark had worked with both men previously when all three served on MacArthur's staff. Once he explained his mission, they were eager to help him. As a precaution should they be captured by the Communists, they agreed that it would be best if the two South Korean officers used aliases, so from that moment on they would call Youn Mr. Yong Chi Ho and Ke Mr. Kim Nam Sun.[9]

Yong was bilingual (Korean, English) and a veteran of guerrilla fighting in China and Korea. He proved to be a loyal and sometimes ruthless warrior. Kim had been a counterintelligence officer in the South Korean navy. He spoke no English, but his interrogations of the Communist soldiers and sympathizers they captured were relentless and effective.

The following day, Clark flew back to Tokyo with Yong and Kim, and they spent the next two days preparing. They tentatively selected the island of Tokchok-do as their base and made lists of the equipment and weapons they would need and the information they planned to gather.

To help him get the supplies he needed, Clark contacted an old shipmate, Lieutenant Commander Russell "Shorty" June, now the executive officer of the Saesbo Naval Base. After hearing of Clark's mission, June instructed his supply officer to provide everything Lieutenant Clark wanted. He got grease guns (submachine guns), thousands of rounds of ammunition, fifty cases of World War II C-rations, a large tent to serve as Clark's command post, a radio to keep in touch with Tokyo, one million won (South Korean currency, worth about US $550 in 1950) in case he needed to buy information, two hundred pounds of rice, dried fish, two cases of Canadian Club to loosen tongues, and nearly everything else Clark wanted except M1 Garand rifles and hand grenades.

A British ship, the *Charity,* had been assigned to take Clark and his lieutenants from Sasebo, Japan, to rendezvous with a South Korean patrol boat in the Yellow Sea for the final run to the Inchon area. The *Charity* took them and their large pile of supplies as far as the mouth of Flying Fish Channel, the main channel to Inchon, where they met the South Korean patrol boat PC-703. Commander Lee of the PC-703 had orders from his admiral to place his ship under Lieutenant Clark's command. It was September 1, two weeks before D-Day.

Commander Lee carried them to Tokchok-do, about twenty-five miles southwest of Inchon, but after a brief inspection, Clark and Yong decided it was too far from Inchon to be an effective headquarters.[10] Lee took them next to Yonghung-do, an island twelve miles south of Inchon. At that time, it was home to more than a thousand loyal South Koreans, most of whom were anxious to rid their country of the oppressors from the north.

Yonghung-do is roughly circular, about four miles in diameter, with several low mountains abutting the northern and western beaches. The coastline is irregular and indented with many bays. A beach along the island's northern shore was a good landing spot.

Lee took the PC-703 as close to the beach as he could, and two junks sailed out from the village. Clark, Yong, and Kim transferred their supplies to the junks and went ashore. The beach was rocky, some six hundred yards long, backed by a four-foot-high scarp. Promontories extended seaward from each end of the beach, providing good observation posts and protection from all but northerly storms. Beyond the scarp, a flat area covered with trees extended inland about thirty yards. Drying rice paddies spread southward from the trees nearly half a mile to the edge of a village. Clark thought this location could be defended against Red infiltrators, and he set up his command post tent under some wide-branching trees to hide it from aerial observation. Yonghung-do became their base.

A narrow channel separated the easternmost point of Yonghung-do from another island, Taebu-do, where the NKPA had a three-hundred-man army garrison. A ferry landing at this point was a likely place for an attack by the Reds. The channel was fordable at the low spring tides during Clark's first two days at Yonghung-do. From then until September 12, the water would be too deep to wade, but after that, the NKPA soldiers could walk across to attack them.

Having enemy soldiers so close was a constant concern.[11] Clark had hardly dropped his sea bag when the village headman told him there were five NKPA soldiers on Yonghung-do. Before unpacking anything, he and Yong armed a group of young men and set off in search of the Inmingun. He hoped to capture and interrogate them, but his overzealous guards shot the Reds as they tried to flee to Taebu-do. This incident convinced Clark that the North Koreans were aware that something unusual was going on, and he knew they would probe until, little by little, they pieced together who was there and what they were doing.

Yong recruited seven young men from the village to serve as camp guards and bodyguards. Youngsters were all that were left. The NKPA had conscripted all men between the ages of about eighteen and fifty to serve their cause. One camp guard, Chae, seemed to appoint himself as Clark's aide. Several dozen more volunteered and were assigned as lookouts to watch from high points and the ferry landing. Clark only had enough machine guns to arm the camp guards and the ferry guards. The best he could do for the lookouts was to tell them to run back to camp if they saw anything unusual. He wished he had thought to bring field radios and telephones.

The schoolmaster brought two young women, Cynn and Lim, and a middle-aged woman, Moon, to serve as their cooks and maids.[12] Lim was the schoolmaster's daughter and spoke good English. They soon put her to work taking notes during Kim's interrogations.

Very few of the islanders other than Lim spoke English; however, most over the age of ten spoke Japanese. Clark was fluent in the language because of his many years in the Orient, so Japanese became the language of choice for conversations with the islanders.

The next morning, Clark and his Korean lieutenants huddled to decide how they would get the information they needed.[13] The first priority was information on the two channels leading to Inchon (Flying Fish Channel and East Channel). A key concern was to determine if there were mines in the channels. Where were navigation aids? One ship disabled in the narrow waterway could block all behind it. They established daily patrols along both channels to search for mines. The second priority was to locate the exact positions of all NKPA shore gun batteries, barracks, and supply dumps at Inchon so they could be destroyed either by aerial or naval bombardment before D-Day. Sending people to Inchon seemed the best way to get this information. The third item was information on Inchon itself: heights of the seawalls, nature and extent of the mudflats fronting the seawalls, exits from the beaches, roads to Seoul. Clark and Yong decided to gather most of this information personally.

That evening, September, 2, the fringes of Typhoon Jane pounded Yonghung-do with heavy rain and wind.[14] It spared them the full force of its 120-knot winds but kept the NKPA pinned down on Taebu-do. Kim took advantage of the lull in the Reds' harassment to question the islanders on NKPA activities. They pinpointed the billets of about a thousand troops in churches and warehouses in Inchon and reported seeing Russian T-34 tanks. By midnight, he had enough useful leads for all their future planning and operations.

The next morning, Clark noticed ten to fifteen boats that had sought refuge from the typhoon near the western promontory. They belonged to fishermen on Yonghung-do and neighboring islands. Yong sought out the chief of the local fishermen's association, a man named Chang who "had all the physical attributes and appearance of a China coast pirate," Clark wrote.[15] After some crafty negotiation, Clark donated five sacks of rice to the association, and Chang and the boatmen agreed to repair and rerig their boats and make them available for use against the Inmingun. Clark appointed Chang commander of the fleet and invited him to make his headquarters at the command post. He planned to use the boats extensively and wanted Chang close at hand. The repairs were made, and by evening Lieutenant Clark's navy consisted of two seagoing junks, three coastal junks, and four sampans.[16]

The sampans were small, flat-bottom boats, fifteen to twenty feet long that were propelled by sail and by sculling with a long oar. They were useful for travel between islands. The junks were larger sailing craft, some thirty-five feet in length, capable of sailing open seas. The fishermen mounted a .50-caliber machine gun in the bow of one of the junks, which served as Chang's "flagship," and .30-caliber machine guns in the others.

In the days that followed, if Clark had no specific job for him, Chang and three of his captains hid among the midchannel islands, waiting for unwary junks. Every day they captured more boats to add to their fleet. If Chang was a pirate, Clark was glad to have him on his side.

~~~

LIEUTENANT CLARK WANTED an alternate base of operations in case the Reds on Taebu-do became too troublesome. The island of Palmi-do, about six miles north of Yonghung-do, seemed like a good candidate. So on September 3, when they planned their first nightly information-gathering foray, Clark and Yong decided to stop at Palmi-do on their way to Inchon to check out the island.

They left the command post at 2000 hours to take advantage of the darkness and the flooding tidal current that would carry them toward Inchon. By 0100, the tide would start to ebb and the moon would rise, so they would have to be on their way back by then. They towed a small sampan to Palmi-do and sent an island boatman named Rhee on to reconnoiter Inchon and, if possible, bring back a motorized junk.
~~~

There was a defunct lighthouse on the small, uninhabited island of Palmi-do. It would make a good base if a move became necessary, so they left three men there with a .30-caliber machine gun.

~~~

LIEUTENANT CLARK BECAME a good student of tides because everything he did had to be coordinated with the rise and fall of the water levels, tidal currents, and hours of moonrise and moonset. Clark wrote, “They [tides] combined to form the key to success if properly used. Conversely, they held the threat of death and disaster if ignored. For the days to come, time would be reckoned not in hourly divisions but by tidal periods. Three hours to flood, three hours to ebb, with moonrise and moonset delimiting our nightly excursions.”[17]

Clark dispatched six sampans to the deepwater channels to patrol for mines. On each change of tide they made a complete run through the channels to look for mines or mining activity. None were found in Flying Fish Channel or in East Channel, but Chang’s boatmen reported seeing NKPA boats laying mines farther to the north, and Clark and Yong knew they would soon be at work in Flying Fish Channel.

As Clark expected, the Inmingun sent soldiers to Yonghung-do. They came at night, and in the morning their beached sampans were discovered, but footprints that disappeared into the hills were the only evidence of the enemy infiltrators. They melted into the hills and became bolder as time went on. Every morning, another dead guard or two confirmed their presence.

Toward the end of his first week on Yonghung-do, Clark decided the infiltrators had to be eliminated.[18] He and Yong led a group of fifty or more villagers in a sweep across the hills. They cornered about twenty Inmingun hiding in a maze of huge rocks at the top of a hill. Below the crest, a steep slope dropped to the sea.

The villagers gathered dry grass and sticks and started a fire among the rocks. Green wood and leaves were added to the blaze to smoke out the enemy, while the guards fired into the smoke. Seven Reds were found dead in a shallow cave. The rest tried to escape to the sea, but foreseeing this, Yong had sent a group to the beach to stop any such move. As they tried to flee, the infiltrators were caught in the open on a steep slope, in a crossfire between a group firing
~~~

at them from above and another firing up from the beach. Seven surrendered. The others joined their ancestors. Later, after intensive interrogation of the infiltrators, Kim remarked that one of the survivors told him they were the advance party for a larger Inmingun invasion force.

With the help of a man captured by the pirate Chang, Clark negotiated with two groups of loyal South Korean guerrillas on the mainland to provide information on the NKPA at Inchon in exchange for rice and guns. Each night they provided a list of targets to Lieutenant Clark, which he radioed to Tokyo after he returned to the command post. The targets were blasted by aircraft the next day. That night the guerrillas would report the damage and have a list of new targets ready for Lieutenant Clark.[19]

ON THE AFTERNOON OF September 7, a lookout brought word that six NKPA junks were headed toward the narrow channel between Yonghung-do and Taebu-do.[20] Clark remembered Kim's comment about an invasion force. This must be it! He called a conference with his lieutenants and Chang, and they decided to attack the NKPA boats—they could not afford to let enemy troops land on Yonghung-do. The tide was at ebb, which favored getting their five junks to sea quickly.

Clark sailed with Chang. He expected Chang to turn east when he cleared the promontory and take the shortest track to intercept the NKPA. Instead, he took a westerly course, which was two miles longer. "The tide is our only friend now, *Taicho-san,*" Chang said. "On this course, we remain in the lee of the island until we come about for the approach on the Inmingun. Should we take the other passage, we will have a much weaker tidal current.... We will save at least thirty minutes by taking the East Channel."[21] Lieutenant Clark appreciated the lesson in local sea lore.

They sailed from the northern shore of Yonghung-do, counterclockwise around the island. A south wind and tide favored them over the Inmingun. Soon they passed the peninsula at the southwest corner of Yonghung-do. Two miles of mudflats separated them from shore. After a little more sailing along the southern coast, they spotted their adversaries—six junks about five miles ahead—coming toward them.

"If we continue on this course, we will get in the tidal eddy ahead and be swept ashore, *Taicho-san,*"[22] Chang explained to Clark as he changed course. When currents meet an obstruction such as an island, the stream divides and flows around it, and eddies form in the lee of the island. The tide had changed on the afternoon of September 7 and was flowing from north to south. Clark's fleet was now on the lee side of Yonghung-do, and he could see the eddy: another lesson in local sea lore well taken.

The "Battle of the Junks" began (see Map 5).

The NKPA fleet was led by what Clark called a Kwangtung, a large junk with an antitank gun on its bow. He estimated some eighty or ninety armed troops crowded the decks of the NKPA armada. After some initial sparring, Clark realized that although the Kwantung's firepower outmatched his, the gun was fixed and could only be aimed by changing the boat's heading. This was an advantage for Clark's fleet with its smaller, swivel-mounted machine guns.

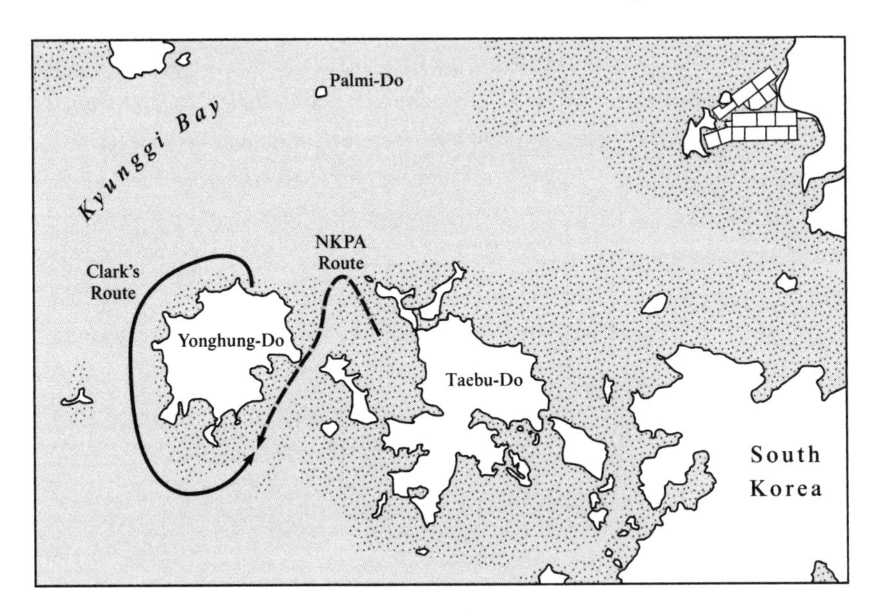

MAP 5

Battle of the Junks, September 7, 1950. The route taken by Lieutenant Eugene F. Clark and his South Korean allies is indicated by the solid line; that of the NKPA, the North Korean People's Army, is indicated by the dotted line. Adapted from University of Texas at Austin Libraries. AMS Topographic Maps, Series L552, US Army Map Service, 1950, Seoul, Korea, NJ-52-9.

The big craft took its toll, though. It sank one of Clark's boats, shot holes through several, including Chang's flagship, and toppled Chang's mainmast. This was a severe blow to a sailing vessel, but Chang merely cursed and sought revenge. Clark positioned his boats abreast of his own, while the NKPA had theirs in line behind the Kwangtung. Clark's navy had the advantage because his boats could all fire simultaneously, concentrating their fire on the lead NKPA boat. Those behind it could not return the fire without hitting each other—disastrous naval tactics.

The South Koreans closed on the NKPA junks, and Clark's gunner, Chae, opened up on the bow of the Kwantung. It was unable to bring its gun to bear on the smaller boat that kept maneuvering out of its sights. Chae blasted away the bow of the Kwangtung. The big gun toppled over backward. Soldiers knocked each other down jumping overboard to avoid Chae's bullets. With the Kwangtung a derelict, Chang attacked the next junk in line, a coaster. Without the Kwangtung's bulk blocking his field of fire, Chae reduced the coaster and its troops to splinters and bodies floating on the tide.

Chang captured two of the NKPA vessels to add to their fleet and sank the others after removing everything of use. His crews shot every NKPA head that appeared above the water's surface. There were no Red survivors.

Clark was aghast at the wanton, unnecessary killing. It was not the way he had been taught to fight a war. He felt compelled to take the soldiers prisoner but had no power to do so. The rules of cooperation between the United States and South Korea gave the South Koreans sole authority over all captured NKPA. Lieutenant Clark was almost thankful that the decision was not his to make. There were too many Inmingun on Yonghung-do already.

Lieutenant Clark had just won the only naval battle of his career.

~~~

THE ISLAND OF WOLMI-DO stands between Inchon and the open sea, in a perfect location to defend the city against an amphibious invasion. The hills above the waterfront are honeycombed with caves. The NKPA hid artillery in the caves and in buildings in the city, and had machine-gun nests and troop bunkers camouflaged along the waterfront. Admiral Doyle, the invasion task force commander, knew Wolmi-do's guns could inflict unacceptable losses on
~~~

the invasion forces if they were not knocked out before D-Day. The problem was that everything was so well hidden that little was visible from the air. This put Wolmi-do at the top of Clark's priority list.

Commander Lee, the captain of the South Korean patrol boat that took Lieutenant Clark to Yonghung-do, had a unique way to locate the guns. He steamed to within a few miles of Wolmi-do (but just out of gun range) and ran across in front of the island, knowing the NKPA gunners couldn't resist this tempting target. When they fired at the PC-703, Commander Lee noted every gun flash on shore. Back at Yonghung-do, he and Clark plotted the gun positions and reported the information to Tokyo. The next morning marine aircraft bombed the hidden guns.

One causeway connected Wolmi-do to the mainland, another to Sowolmi-do, a smaller island. Seawalls along the Inchon waterfront would be significant obstacles to the invading marines. The planners in Tokyo had to know how high the Inchon seawalls were. Fortunately, there were no seawalls at Wolmi-do.

Clark and Yong made two separate reconnaissance trips: the first to Wolmi-do and the second to Inchon.

Clark, Yong, and Chae went to Wolmi-do on an incoming tide. Chae went ashore on the northern end of the island, while their boatman took Yong and Clark farther south to the causeway between Wolmi-do and Sowolmi-do. The causeway held no guns, but there were several machine-gun nests at its juncture with Sowolmi-do. The little island was clear. They worked their way back to the rendezvous point and waited for Chae. The tide had changed and carried them home at three to four knots.

Back at the command post, they debriefed Chae. A cove on the northwestern side of Wolmi-do was clear and would make a good landing beach. It became Green Beach on D-Day. A hard-surfaced road circled the western end of the waterfront and joined the causeway, which then went on to the main road to Seoul. Trenches had been dug on the landward side of the seawall and were fronted by high barbed-wire fences. Old charts showed a ramp up from the beach, but it was no longer there. This was all important information for Tokyo.

That afternoon, Rhee, the boatman they had sent to Inchon to find a power boat, returned with a pompom, a thirty-five-foot junk powered by a one-cylinder diesel engine. What a difference it would have made in the previous day's battle, but Clark put it to good use the next week. He was no longer at the mercy of the tides. Rhee had hardly anchored when Clark and Yong appeared in the wheelhouse and asked him to take them to Palmi-do.[23] If they could light the beacon of the defunct lighthouse, it would be of immeasurable value to General MacArthur's ships as they traveled the winding Flying Fish Channel at night.

The batteries that powered the beacon's rotating shade had been destroyed, but the light itself was a large kerosene lantern, and there was even a half-full can of kerosene nearby. Clark lit the lamp briefly to determine that it worked and then extinguished it. He could provide a steady light to guide the ships up Flying Fish Channel. When they left, Clark could not resist writing "Kilroy Was Here" on the lighthouse door. They returned to the command post and decided that conditions that night were favorable for reconnoitering the Inchon beaches: the moon was dark, and the tide would start to flood at 2050 hours, creating currents to carry them north.

Because the tide was out and exposed the mudflats, the boatman could get them no closer than five hundred yards from the seawall. When Clark stepped over the side of the sampan, "It was like stepping into a big can of grease,"[24] he wrote. It took half an hour to wade to the seawall, leaving them completely exhausted. Clark quickly decided that neither men nor vehicles could negotiate the soft goo, and another of his questions was answered.

With that important information to pass on to Tokyo, and having finally reached shore, Clark and Yong began their reconnaissance of the Inchon waterfront. They found that a road skirting the seawall lead into the Inchon suburbs and on to the highway to Seoul. There was a break in the seawall that bulldozers could widen enough for vehicles to pass through. Dry rice paddies behind the wall would support vehicle traffic. The tide was the only problem. Large landing craft could not get close to the seawall, but smaller boats could be used for two or three hours at high tide. This beach on the south side of Inchon harbor would be designated Blue Beach by the planners in Tokyo. They looked for other favorable landing sites south of the city but found none.

Their last task was to measure the height of the seawall.[25] Clark climbed as high as he could and braced himself against the wall. Yong climbed up on his shoulders, stretched his arms to their full length, and his outstretched fingers just reached the top of the wall. When they got back to the command post, the men measured their combined height and determined that the seawalls were fourteen to sixteen feet high—too high to climb over from a beached landing craft. Scaling ladders would have to be built in Japan.

The next day, the Tokyo planners asked about the accuracy of the American and Japanese tide tables they were using. Clark told them the Japanese tables were more accurate.[26] The planners asked him to turn the lighthouse light on at 0030 hours on the morning of September 15 and told him that US and British cruisers would begin shelling Inchon and Wolmi-do the next morning as part of the softening up for D-Day.

~~~

ABOUT NOON ON September 14, Clark's last day of surveillance, a young ferry guard limped into camp. Infiltrators had cut his guard group off from the village. The intrepid youth had escaped by swimming out into the channel and letting the tidal current carry him past the Inmingun.[27]

It looked like the Reds would attack that afternoon, and the invasion fleet would not arrive until the following morning. Clark set up a mile-long defensive line south of the village. He knew they could not hold long against a coordinated assault, but he hoped the defenders could warn the villagers at the command post when the attack came. His only option now was to evacuate himself, Yong, and Kim and as many of the villagers who had helped him as possible. Clark expected an attack at dusk. The guards along the defensive perimeter had orders to fight but fall back to the beach when they could hold out no longer.

All was still quiet at six o'clock, but to be safe, he ordered the villagers aboard the junks. Only about three hundred could be persuaded to leave. A guard brought word that four or five junks had left Taebu-do. Clark figured it would take them about two hours to reach the south side of Yonghung-do and get to the village. All was dark now. Chae had a .50-caliber machine gun in a defensive position between the beach and the rice paddies.

About ten thirty, firing from beyond the village indicated the Reds had landed and were approaching. The guards came in, and the Inmingun quickly controlled the village. Dark forms materialized in the rice paddies and advanced toward the beach.

More junks rounded the east promontory—a two-pronged attack! Clark sent
~~~

Chae to the beach with his machine gun to do as much damage to the junks as he could. He sent a message to Kim for their junks to shove off.

The guards fought a delaying action as long as they could but finally had to abandon their defenses and run to the boats. Clark crossed the mudflat to the west promontory and was nearly shot by his own guards before he remembered to shout the password. He scrambled across the rocks to a sampan that took him to the pompom. The old diesel was roaring, spewing sparks ten feet into the air. Clark clambered aboard and found Yong and Kim already there. Rhee set a course for Palmi-do.

When they reached the island, they dashed to the lighthouse.[28] Clark was so exhausted when he got there he could hardly climb the steel ladder to the top, but somehow he did and lit the kerosene beacon at 0050. It distressed him immensely that he was twenty minutes late lighting the lighthouse light. Then he passed out on the catwalk around the light.

Yong woke Clark at the first light of dawn, and he was astounded by the scene spread out below them. From their perch more than two hundred feet above the water, they saw 230 ghostly ships and many smaller craft that had glided in during the night, poised to attack Inchon (see Map 6).

Clark wondered if he had done all he could, if the attack would be successful, and what would happen to the islanders who had helped him. The pompom took him and his two Korean officers (Yong and Kim) to MacArthur's flagship, the USS *Mount McKinley*. They had a ringside seat at the invasion to which they contributed so much.

Lieutenant Clark was concerned about his friends on Yonghung-do, so as soon as he arrived on the *Mount McKinley*, he tried to persuade the officers in charge to rescue his allies there, but they could not alter their timetable for the invasion until the next day. A battalion of marines was sent to sweep the North Koreans off Yonghung-do and Taebu-do, which they did, but not before the NKPA murdered more than fifty people who had helped Clark.

When the ships reached the Inchon area on the morning of September 15, a battalion of Lieutenant Colonel Robert D. Taplett's 3rd Battalion, 5th Marines transferred to seventeen LCVP landing craft (Higgins boats), the same as were used at Normandy,[29] and nine Pershing tanks were loaded into logistic support vessels. At 0633 hours, September 15, 1950, the first wave of marines, supported by tanks and naval gunfire, landed on the morning high spring tide at Green

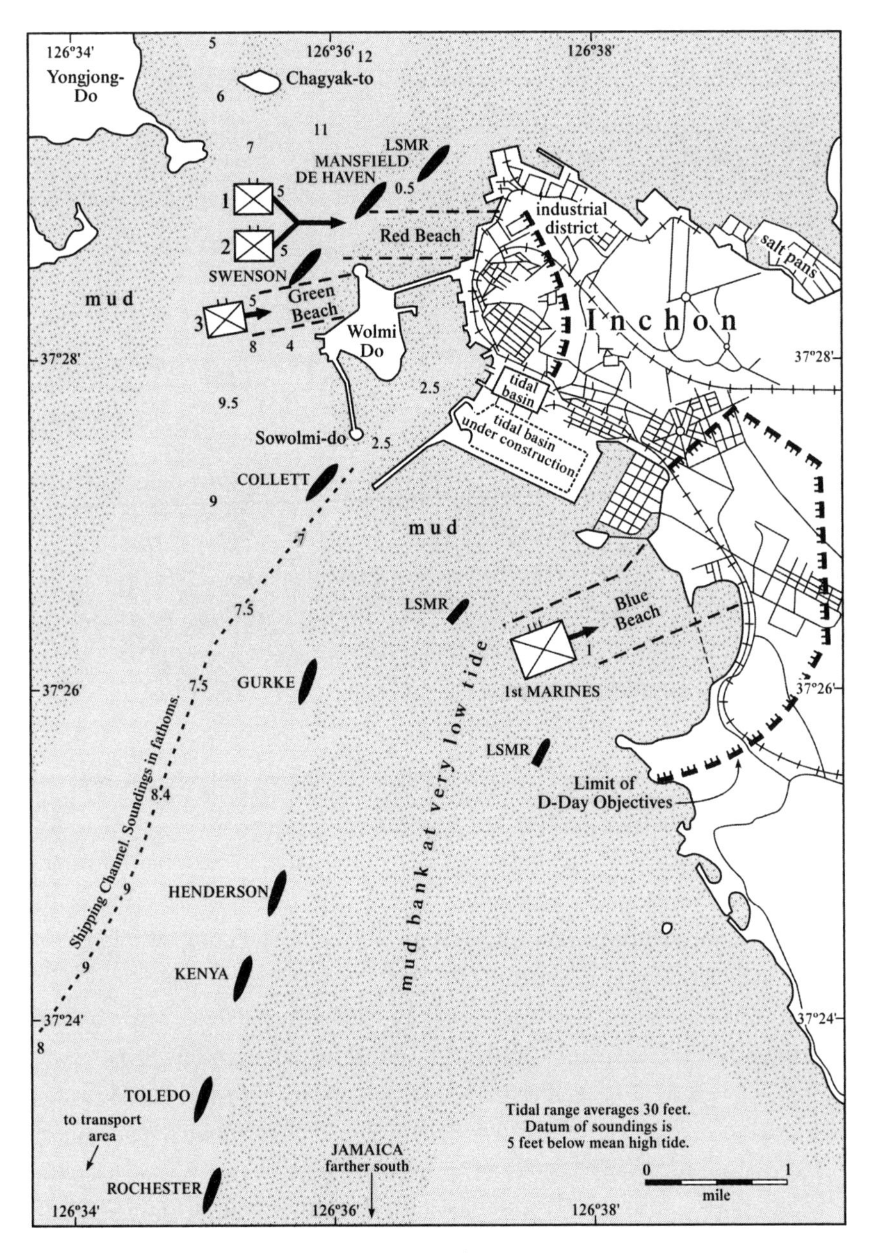

MAP 6

Map of Inchon, September 15, 1950, showing positions of some of the assault ships. Note the extensive mudflats exposed at low spring tides and the causeways connecting Wolmi-do to Sowolmi-do and to the mainland.

FIGURE 21

LSTs (landing ship, tank) unload on Red Beach at Inchon, September 16, 1950.

Beach on Wolmi-do. Once ashore, the marines formed up into their units and attacked the NKPA artillery hidden in the hills. By noon they had captured the island and formed defensive positions on Wolmi-do; then they waited for the evening high tide at 1730 hours.

At 1530, the second wave of marines, the main invasion force, began unloading from their transport ships into LCVPs. These assault troops from the 1st and 5th Marines landed on Red Beach at Inchon, and the marines quickly scaled the seawalls using ladders that had been specially built in Japan based on the measurements made by Yong and Lieutenant Clark.

As soon as the assault troops were safely lodged ashore, eight specially loaded LSTs landed on Red Beach (see Figure 21). With a high spring tide, the LSTs, which required water twenty-nine feet deep, ran up to the beach and grounded. The large ships were stranded on the mudflats and the marines spent the night unloading vehicles and supplies for the next day's operations.

On the morning high spring tide the next day, September 16, another wave of LSTs switched places with those on Red Beach. They unloaded during the day and switched places with the next incoming wave of LSTs on the evening high tide. As was the case with Lieutenant Clark, proper use of the tides held the key to success or disaster for the large LSTs and their cargoes of men and equipment.

The marines killed or captured all the NKPA soldiers on Wolmi-do during the night of September 15. By September 18, the marines had captured Inchon and Kimpo airfield, and aircraft could operate in close support of the Allied troops as they advanced on Seoul. The marines quickly reached Seoul and the city was in UN hands by September 26. General MacArthur's gamble paid off handsomely.

Lieutenant Clark's mission has been called "the most daring covert mission of the Korean War."[30] He retired from the navy in 1966 with the rank of commander and lived in California and Nevada with his wife until his death in 1998 at the age of eighty-six.

AFTERWORD

MANY OF THE TRAINED WEATHERMEN and oceanographers who contributed to the success of innumerable amphibious and terrestrial operations during World War II and the Korean War mustered out of the military after those conflicts. Much of their expertise was lost to the military, but they went on to play vital roles in academia, government service, and private industry in developing the present state of the art of meteorology and oceanography in the United States.

Dr. Harald Sverdrup and Dr. Walter Munk wrote a classified report of their work in 1943, and in March 1947, the navy's Hydrographic Office released an unclassified, slightly revised version of the report, *Wind, Sea and Swell: Theory of Relations for Forecasting* The most important revision was the introduction of the concept of "significant waves," which they defined as "waves having average height and period of the one-third highest waves."[1] After leaving active duty, Al H. Glenn recognized that the oil companies needed scientifically developed oceanographic and meteorological design criteria for their new Gulf of Mexico offshore drilling facilities. He formed A. H. Glenn and Associates in 1946 to provide these criteria to clients in the oil industry. Glenn used the principles set forth in *Wind, Sea and Swell* as the basis for his criteria development until a more advanced spectral ocean wave model system developed by Oceanweather, Inc., began to supplant it in the late 1970s. A. H. Glenn and Associates was based in Grand Isle, Louisiana, and then in New Orleans. Charles C. Bates worked with Glenn for a while after he left active duty in the army.

Offshore oil drilling and production systems had to be designed to survive the most extreme environmental conditions that could reasonably be expected to occur in the area in which they were to be deployed. Detailed weather and ocean response forecasts were needed for these fixed facilities, which could not flee to protected waters as the shrimp boats could with the approach of a tropical storm or hurricane. The design engineers needed to know the most extreme storm wave heights and the directions from which they would approach, extreme current speeds and directions, highest wind speeds and directions, extreme air and water temperatures to be expected, and a host of other environmental parameters. Wind and current speeds and directions change with height and depth below the surface, and this is important to some design problems.

This was an opportunity for the specially trained meteorological oceanographers of the army's program at the University of Southern California and Scripps Institution of Oceanography. Most of these highly educated military officers were now out of the service and were either looking for employment or had entered graduate school. The oil companies' needs presented opportunities for some of the entrepreneurially inclined among these recent servicemen. Paul L. Horrer, another student of Sverdrup and Munk in the army program, enrolled in the Scripps Institution as a graduate student, class of 1946, after the war and then founded Marine Advisors in La Jolla, California, in 1955. It later became Intersea Research Corporation and provided design criteria and other services to the oil industry and government clients.

Many other army meteorologists and oceanographers who trained at USC and Scripps went on to illustrious careers in academia after leaving military service. Dale F. Leipper founded the Department of Oceanography at Texas A&M University in 1949, and a similar department at the US Naval Postgraduate School in 1968.[2] Bob Reid accompanied Dale Leipper to Texas A&M as one of the first physical oceanographers on Texas A&M's staff. Charles Bates left the employ of Glenn and Associates and matriculated at Texas A&M, where he obtained his PhD in physical oceanography. As a civilian, Dr. Bates continued to serve his country; he held key positions in the Department of Defense's Advanced Research Projects Agency, was technical director of the US Naval Oceanographic Office, and was science advisor to the commandant of the US Coast Guard. In this latter position Dr. Bates initiated a US Navy program for observing and forecasting ice conditions in the American arctic.[3]

Warren C. Thompson, a US Navy ensign, had been a high school classmate and close friend of Bob Reid before World War II. He also was a graduate of the Sverdrup/Munk oceanography program and attended Texas A&M after the war to obtain a PhD in physical oceanography. Dr. Thompson joined the Naval Postgraduate School in Monterrey, California, in 1953 where he helped initiate the program in oceanography. Donald W. Pritchard, who commanded Weather Detachment YK at Normandy, returned to Scripps Institution of Oceanography after the war and earned master's and PhD degrees in oceanography. Dr. Pritchard then took the position as the first head of Johns Hopkins University's Chesapeake Bay Institute in 1949 and a year later founded Johns Hopkins's oceanography department. John Crowell left the army as a captain at the end of the war and earned a master's degree in oceanographic meteorology from Scripps in 1946 and a PhD in geology from UCLA in 1947. He left there twenty years later as a distinguished professor and went to the University of California, Santa Barbara, and cofounded the Environmental Studies Program. Dr. Crowell's work was recognized with many awards, as well as membership in the National Academy of Sciences, a fellowship in the American Academy of Arts and Sciences, a Fulbright Award, a Guggenheim Scholarship, and several other honorary degrees and awards.

These are but a few of the many military officers who trained in the meteorological/oceanographic program at the University of Southern California and Scripps Institution of Oceanography during World War II who went on to distinguished careers as meteorologists and oceanographers and taught and inspired many others in these fields, the present author included.

The type of beach forecasting that the Swell Forecast Section did for Normandy, and that other forecasters did for other invasion beaches during World War II and at Inchon, ended after the Korean War because of technological advances and the changing nature of warfare. The army and Marine Corps continued research and planning for across-the-beach emplacement of troops on hostile foreign shores but not in the "traditional" style of past amphibious landings where large naval vessels stood off the invasion beach and shelled it with high explosive shells while assault troops stormed toward the beach in landing craft and then waded the last few hundred yards through the surf and enemy gunfire to reach dry ground. Instead, large ships have been designed to carry helicopters and MV-22 Osprey tilt-rotor aircraft to ferry marines and soldiers over the shore and into battle, enabling troops to land behind enemy lines and avoid the beach and surf zone entirely. In addition, the enemy ashore can be expected to have guided missiles that can strike the naval troops and fire at support vessels long before they are within range of the beach, making traditional landing assaults obsolete.

The army and the other military services gained access to many more weather and ocean forecasting resources than General Eisenhower had for the D-Day invasion at Normandy. Research by the army, the navy, the National Weather Service (which includes the National Hurricane Center), universities, and private weather services have continually upgraded state-of-the-art weather and ocean forecasting. Now daily forecasts provided to the public are based on far more sophisticated and exacting models than anything General Eisenhower's meteorologists and oceanographers had available for what many authors and historians have called the most important weather forecast in history.

NOTES

Chapter One

1. Robert D. Paulus, "Pack Mules and Surfboats: Logistics in the Mexican War," *Army Logistician* 29, no. 6 (1997): 34–40; K. Jack Bauer, *Surfboats and Horse Marines: U.S. Naval Operations in the Mexican War, 1846–48* (Annapolis, MD: US Naval Institute, 1969), 66.

2. Bauer, *Surfboats and Horse Marines,* 77.

3. Hans E. Rosenthal, "Northers of the Gulf of Mexico and Central American Waters," *Mariners Weather Log,* November 1965; Jan Reid, "Texas Primer: The Blue Norther," *Texas Monthly,* November 1982, https://www.texasmonthly.com/the-culture/texas-primer-the-blue-norther/.

4. Bauer, *Surfboats and Horse Marines,* 82.

5. Bauer, *Surfboats and Horse Marines,* 85.

6. Major Jeffrey L. Hammond, USMC, *Marine Corps Interwar Period Innovation and Implications for the Upcoming Post Operation Enduring Freedom Period,*_ http://www.dtic.mil/dtic/tr/fulltext/u2/a583991.pdf, 9; Edwin Howard Simmons, "Amphibious Warfare," in *The Oxford Companion to American Military History,* ed. John Whiteclay Chambers II et al. (Oxford: Oxford University Press, 1999), 31–32.

7. Hammond, *Marine Corps Interwar Period Innovation,* 9; Allan R. Millett, "Assault from the Sea: The Development of Amphibious Warfare between the Wars—The American, British, and Japanese Experiences," in *Military Innovations in the Interwar Period,* ed. Williamson Murray and Allan R. Millett (Cambridge: Cambridge University Press, 1996), 72.

8. Earl H. Ellis, USMC, *Advanced Base Operations in Micronesia,* HyperWar: A Hypertext History of the Second World War, http://www.ibiblio.org/hyperwar/USMC/ref/AdvBaseOps/index.html, 31.

9. Hammond, *Marine Corps Interwar Period Innovation,* 11.

10. Jerry E. Strahan, *Andrew Jackson Higgins and the Boats That Won World War II* (Baton Rouge: Louisiana State University Press, 1994), 24.

11. Strahan, *Andrew Jackson Higgins,* 27.

12. Strahan, *Andrew Jackson Higgins,* 48.

13. Strahan, *Andrew Jackson Higgins,* 48, 88.

14. Hammond, *Marine Corps Interwar Period Innovation,* 12.

15. *Beaching the L.S.T.,* LST training video, US Atlantic Fleet Amphibious Force Landing Craft Group, Photographic Science Laboratory, Naval Air Support Facility Anacostia, Washington, DC, YouTube, https://www.youtube.com/watch?v=9eCCgRCzN1Y.

16. Charles C. Bates and John F. Fuller, *America's Weather Warriors, 1814–1985* (College Station: Texas A&M University Press, 1986), 52.

Chapter Two

1. Bates and Fuller, *America's Weather Warriors,* 46.

2. C. V. Glines, "Airmail Service: It Began with Army Air Service Pilots," HistoryNet, 12 June 2006, http://www.historynet.com/airmail-service-it-began-with-army-air-service-pilots.htm, originally published in *Aviation History,* May 1994.

3. National Oceanic and Atmospheric Administration, "NOAA Celebrates 200 Years of Science, Service, and Stewardship," https://celebrating200years.noaa.gov/foundations/welcome.html.

4. Bates and Fuller, *America's Weather Warriors,* 45.

5. Bates and Fuller, *America's Weather Warriors,* 46.

6. Bates and Fuller, *America's Weather Warriors,* 52.

7. Robert O. Reid, "Robert Osborne Reid Memoirs," vol. 1, "The Formative Years: 1921–1951," unpublished manuscript, 2008, 53.

8. Bates and Fuller, *America's Weather Warriors,* 276.

9. Reid, "Memoirs," 55.

10. Claudia Geib, "How a Father and Son Helped Create Weather Forecasting as We Know It," Atlas Obscura, 22 November 2016,https://www.atlasobscura.com/articles/how-a-father-and-son-helped-create-weather-forecasting-as-we-know-it.

11. Geib, "How a Father and Son Helped Create Weather Forecasting."

12. Reid, "Memoirs," 56.

13. William A. Nierenberg, "Harald Ulrik Sverdrup, 1888–1957," National Academy of Sciences, Biographical Memoirs, http://www.nasonline.org/publications/biographical-memoirs/memoir-pdfs/sverdrup-harald.pdf

14. "Beaufort Wind Scale," Stormfax Weather Almanac, http://www.stormfax.com/beaufort.htm.

15. Bates and Fuller, *America's Weather Warriors,* 56.

16. Walter Munk and Deborah Day, "Harald U. Sverdrup and the War Years." *Oceanography* 15, no. 4 (2002): 22.

17. Charles C. Bates, "Sea, Swell, and Surf Forecasting for Operation Overlord," in *Some Meteorological Aspects of the D-Day Invasion of Europe, 6 June 1944,* ed. R. H. Shaw and W. Innes (Boston: American Meteorological Society, 1986), paper originally presented during symposium proceedings, Fort Ord, CA, 19 May 1984, 36.

18. Harald U. Sverdrup and Walter H. Munk, *Wind, Sea and Swell: Theory of Relations for Forecasting.* (U.S. Navy Dept., Hydrographic Office, pub. no. 601, 1947), 2.

19. Munk and Day, "Harald U. Sverdrup and the War Years," 23.

20. Blair Kinsman, *Wind Waves: Their Generation and Propagation on the Ocean Surface* (Englewood Cliffs, NJ: Prentice Hall, 1965), 321.

21. Munk and Day, "Harald U. Sverdrup and the War Years," 16.

22. John C. Crowell, *Surf Forecasting for Invasions during World War II* (Santa Cruz, CA: Marty Magic Books, 2010), 37.

23. John C. Crowell, interview with author, 12 September 2013.

24. Crowell, *Surf Forecasting,* 43.

25. Reid, "Memoirs," 56.

Chapter Three

1. US Army Center of Military History, History of COSSAC, http://www.history.army.mil/documents/cossac/Cossac.htm, 1; James M. Stagg, *Forecast for Overlord, June 6, 1944* (New York: W. W. Norton, 1971), 10–16.

2. Stagg, *Forecast for Overlord,* 13.

3. Stagg, *Forecast for Overlord,* 15.

4. Crowell, *Surf Forecasting,* 45.

5. Crowell, *Surf Forecasting,* 46.

6. Crowell, *Surf Forecasting,* 46.

7. Crowell, *Surf Forecasting,* 52.

8. Charles C. Bates, email to author, 9 January 2010.

9. Bates, "Sea, Swell, and Surf Forecasting for Operation Overlord," 33.

10. Bates, "Sea, Swell, and Surf Forecasting for Operation Overlord," 34.

11. Bates, "Sea, Swell, and Surf Forecasting for Operation Overlord," 34; Crowell, *Surf Forecasting,* 60.

12. Bates, "Sea, Swell, and Surf Forecasting for Operation Overlord," 34.

13. John C. Crowell, "Sea, Swell, and Surf Forecasting Methods Employed for the Invasion of Normandy, July 1944," (master's thesis, University of California at Los Angeles, 1946).

14. John C. Crowell, "Sea, Swell, and Surf Forecasting Methods Employed for the Invasion of Normandy, July 1944," (master's thesis, University of California at Los Angeles, 1946), 58; and Bates, "Sea, Swell, and Surf Forecasting for Operation Overlord," 35.

15. Charles C. Bates, "Sea, Swell and Surf Forecasting for D-Day and Beyond," unpublished manuscript, 2010, 12; Crowell, *Surf Forecasting,* 63.

16. Stagg, *Forecast for Overlord,* 29.

17. Stagg, *Forecast for Overlord,* 46.

18. Stagg, *Forecast for Overlord,* 47.

19. Stagg, *Forecast for Overlord,* 40.

20. Stagg, *Forecast for Overlord,* 45.

21. Stagg, *Forecast for Overlord,* 76.

22. Stagg, *Forecast for Overlord,* 81.

23. Stagg, *Forecast for Overlord,* 102.

24. Stagg, *Forecast for Overlord,* 112.

25. C. K. M. Douglas, "Forecasting for the D-Day Landings," *Meteorological Magazine* 81 (1952): 20.

26. Irving P. Krick, "Role of Caltech Meteorology in the D-Day Forecast," in *Some Meteorological Aspects of the D-Day Invasion of Europe, 6 June 1944,* ed. R. H. Shaw and W. Innes (Boston: American Meteorological Society, 1986), paper originally presented during symposium proceedings, Fort Ord, CA, 19 May 1984, 25.

27. Charles C. Bates, email to author, 28 January 2010.

28. John Ross, *The Forecast for D-Day and the Weatherman behind Ike's Greatest Gamble* (Guilford, CT: Lyons, 2014); Stagg, *Forecast for Overlord,* 125; Bates and Fuller, *America's Weather Warriors,* 95, 259.

29. Ross, *The Forecast for D-Day,* 193.

30. Stagg, *Forecast for Overlord,* 126.

31. Stagg, *Forecast for Overlord,* 126.

Chapter Four

1. Reid, "Memoirs," 71.

2. "Loose Lips Sink Ships," Unifying a Nation: World War II Posters from the New Hampshire State Library, https://www.nh.gov/nhsl/ww2/loose.html.

3. Charles C. Bates, email to author, 30 January 2010.

4. Reid, "Memoirs," 75.

5. A. T. Doodson and H. D. Warburg, *Admiralty Manual of Tides* (London: Her Majesty's Stationery Office, 1952).

6. "Burntcoat Head Park," Atlas Obscura, https://www.atlasobscura.com/places/burntcoat-head-park.

7. NOAA, *Our Restless Tides,* https://tidesandcurrents.noaa.gov/restles3.html.

8. Bruce Parker, "The Tide Predictions for D-Day," *Physics Today* 64, no. 9 (1 September 2011), https://doi.org/10.1063/PT.3.1257.

9. Robert O. Reid, interview with author, 9 August 2006.

10. Reid, "Memoirs," 75.

11. Combined Operations, "Mulberry Harbours," http://www.combinedops.com/Mulberry%20Harbours.htm, "D-Day+."

12. Combined Operations, "Mulberry Harbours," "Final Specification."

13. 25th US Naval Construction Regiment, "Action Report for June 6 to 22, 1944," http://www.seabeecook.com/history/25th_ncr/action_report/action_report_body.htm

14. John C. Crowell, email to author, 19 March 2009.

15. John Fleming, "Recollections of Weather Forecasts Issued for the Normandy Landings in June, 1944," letter to Dr. Charles C. Bates.

16. Combined Operations, "Mulberry Harbours," "D-Day+."

17. 25th US Naval Construction Regiment, "Action Report for June 6 to 22, 1944."

18. Combined Operations, "Mulberry Harbours," "D-Day+."

19. Bates and Fuller, *America's Weather Warriors,* 97.

20. Reid, "Memoirs," 79.

21. Rick Atkinson, *The Guns at Last Light: The War in Western Europe, 1944–1945* (New York: Henry Holt, 2013), 120.

22. Roland G. Ruppenthal, *Logistical Support of the Armies,* vol. 2, *September 1943–May 1945,* United States Army in World War II, The European Theater of Operations, (Washington, DC: Center of Military History, United States Army, 1995), 53.

23. Ruppenthal, *Logistical Support of the Armies,* 2:57.

24. Bates and Fuller, *America's Weather Warriors,* 97.

25. Reid, "Memoirs," 81.

26. Reid, "Memoirs," 88, 89.

Chapter Five

1. Gerhard Neumann and Willard J. Pierson Jr. *Principles of Physical Oceanography* (Englewood Cliffs, NJ: Prentice Hall, 1966), 359.

2. "Command History: Seventh Amphibious Force," part I, HyperWar: A Hypertext History of the Second World War, http://www.ibiblio.org/hyperwar/USN/Admin-Hist/OA/419-7thAmphib/7thAmphibs-1.html, I-14.

3. Major General Hugh J. Casey, *Engineers of the Southwest Pacific, 1941–1945,* vol. 4, *Amphibian Engineer Operations* (Washington, DC: Government Printing Office, 1959), 545.

4. Casey, *Engineers of the Southwest Pacific,* 4:677.

5. Casey, *Engineers of the Southwest Pacific,* 4:492.

6. Casey, *Engineers of the Southwest Pacific,* 4:481.

7. Major General Hugh J. Casey, *Engineers of the Southwest Pacific 1941–1945,* vol. 8, *Critique* (Washington, DC: Government Printing Office, 1959), 491.

8. Chris Schaefer, *Bataan Diary: An American Family in World War II, 1941–1945.* (Houston: Riverview, 2004), 20.

9. Trumbull Higgins, *Korea and the Fall of MacArthur: A Précis in Limited War* (New York: Oxford University Press, 1960), 44.

10. David Rees, *Korea: The Limited War* (Baltimore: Penguin Books, 1964), 80.

11. "Operation Mincemeat—As Told by Ewen Montagu: The Operation Orders," in *The Man Who Never Was: The True Story of Glyndwr Michael,* http://www.themanwhoneverwas.com/operationmincemeat.html.

12. *The Man Who Never Was: The True Story of Glyndwr Michael*, http://www.themanwhonever was.com/.

13. Casey, *Engineers of the Southwest Pacific*, 8:213.

14. Casey, *Engineers of the Southwest Pacific*, 8:333.

15. Casey, *Engineers of the Southwest Pacific*, 4:552.

16. Casey, *Engineers of the Southwest Pacific*, 4:470.

17. Casey, *Engineers of the Southwest Pacific*, 4:599.

18. Albert N. Garland and Howard M. Smyth, *The Mediterranean Theater of Operations, Sicily and the Surrender of Italy*, United States Army in World War II, Center of Military History (Washington, DC: Government Printing Office, 1965).

19. Casey, *Engineers of the Southwest Pacific*, 4:85.

20. Casey, *Engineers of the Southwest Pacific*, 4:561.

21. Casey, *Engineers of the Southwest Pacific*, 4:165.

22. Casey, *Engineers of the Southwest Pacific*, 4:437.

23. Casey, *Engineers of the Southwest Pacific*, 4:485.

24. Rees, *Korea*, 84.

25. Mary-Louise Quinn, *The History of the Beach Erosion Board, U.S. Army Corps of Engineers, 1930–63*, Miscellaneous Report 77-9 (Fort Belvoir, VA: US Army Corps of Engineers, 1977), 43.

26. Quinn, *History of the Beach Erosion Board*, 44.

27. Telephone interviews by author with Professor. W. C. Krumbein (17 August 1978) and Dr. Martin A. Mason (18 August 1978).

28. Telephone interview by author with Garbis H. Keulegan (17 August 1978).

29. A. M. Kamel and D. D. Davidson, *Hydraulic Characteristics of Mobile Breakwaters Composed of Tires or Spheres*, Technical Report H-68-2, (Vicksburg, MS: US Army Engineer Waterways Experiment Station, 1968), 1.

30. Quinn, *History of the Beach Erosion Board*, 44.

31. Quinn, *History of the Beach Erosion Board*, 45; telephone interview between Major Thomas M. Mitchell and Dr. Martin A. Mason (18 August 1978).

32. Quinn, *History of the Beach Erosion Board*, 45.

33. Casey *Engineers of the Southwest Pacific*, 4:677.

34. Quinn, *History of the Beach Erosion Board*, 48.

35. Telephone interview between Major Thomas M. Mitchell and Dr. Martin A. Mason (18 August 1978).

Chapter Six

1. John Wukovits, *One Square Mile of Hell: The Battle for Tarawa* (New York: New American Library Caliber, 2006).

2. Captain James R. Stockman, USMC, *The Battle for Tarawa*, USMC Historical Monograph, chapter 1, http://www.ibiblio.org/hyperwar/USMC/USMC-M-Tarawa/USMC-M-Tarawa-1.html,

5, appendix B, "Marine Casualties," http://www.ibiblio.org/hyperwar/USMC/USMC-M-Tarawa/USMC-M-Tarawa-B.html, and appendix G, "Brief History of the Gilbert Islands before the Japanese Invasion," http://www.ibiblio.org/hyperwar/USMC/USMC-M-Tarawa/USMC-M-Tarawa-G.html.

3. Wukovits, *One Square Mile of Hell.*

4. Robert Louis Stevenson, *In the South Seas,* web edition published by eBooks@Adelaide, University of Adelaide, South Australia, 2010, http://ebooks.adelaide.edu.au/s/stevenson/robert_louis/s848so/index.html.

5. Wukovits, *One Square Mile of Hell,* 74.

6. George C. Dyer, *The Amphibians Came to Conquer,* chapter 18, "That Real Toughie—Tarawa," http://www.ibiblio.org/hyperwar/USN/ACTC/actc-18.html, 717.

7. Dyer, *The Amphibians Came to Conquer,* 718.

8. Australian Government Bureau of Meteorology, "Dodge Tide," http://www.bom.gov.au/oceanography/projects/ntc/dodge/dodge.shtml.

9. Jeter A. Isely and Philip A. Crowl, *The U.S. Marines and Amphibious War: Its Theory, and Its Practice in the Pacific* (Princeton, NJ: Princeton University Press, 1951).

10. Wukovits, *One Square Mile of Hell,* 67; Stockman, *The Battle for Tarawa,* chapter 1, 4.

11. Dyer, *The Amphibians Came to Conquer,* 703.

12. Isely and Crowl, *The U.S. Marines and Amphibious War.*

13. Wukovits, *One Square Mile of Hell,* 112.

14. Dyer, *The Amphibians Came to Conquer,* 651.

15. Wukovits, *One Square Mile of Hell,* 137.

16. Dyer, *The Amphibians Came to Conquer,* 725.

17. Wukovits, *One Square Mile of Hell,* 121.

18. Wukovits, *One Square Mile of Hell,* 193.

19. Colonel Joseph H. Alexander, USMC, *Across the Reef: The Marine Assault of Tarawa,* National Parks Service, Marines in World War II Commemorative Series, https://www.nps.gov/parkhistory/online_books/npswapa/extContent/usmc/pcn-190-003120-00/sec8.htm.

20. Wukovits, *One Square Mile of Hell,* 216.

21. Dyer, *The Amphibians Came to Conquer,* 705.

22. Dyer, *The Amphibians Came to Conquer,* 717–22.

23. Philip A. Crowl and Edmund G. Love, *U.S. Army in World War II, The War in the Pacific: Seizure of the Gilberts and Marshalls,* chapter 10, http://www.ibiblio.org/hyperwar/USA/USA-P-Gilberts/USA-P-Gilberts-10.html.

24. Alexander, *Across the Reef.*

Chapter Seven

1. Clay Blair, *The Forgotten War: America in Korea, 1950–1953* (New York: Time Books, 1987), 219–36.

2. Stockman, *The Battle for Tarawa,* 72.

3. Eugene Franklin Clark, *The Secrets of Inchon: The Untold Story of the Most Daring Covert Mission of the Korean War* (New York: G. P. Putman's Sons, 2002), 9.

4. Robert Debs Heinl Jr., *Victory at High Tide: The Inchon-Seoul Campaign* (Annapolis, MD: Nautical and Aviation Publishing Company of America, 1980), 40.

5. Heinl, *Victory at High Tide,* 24.

6. Appleman, Roy E. *South to the Naktong, North to the Yalu (June-November 1950)* (Washington, DC: Center of Military History, United States Army, 1992), 489.

7. Heinl, *Victory at High Tide,* 44.

8. Clark, *The Secrets of Inchon,* 8.

9. Clark, *The Secrets of Inchon,* 18.

10. Clark, *The Secrets of Inchon,* 20.

11. Clark, *The Secrets of Inchon,* 41.

12. Clark, *The Secrets of Inchon,* 52.

13. Clark, *The Secrets of Inchon,* 63.

14. Clark, *The Secrets of Inchon,* 81.

15. Clark, *The Secrets of Inchon,* 90.

16. Clark, *The Secrets of Inchon,* 92.

17. Clark, *The Secrets of Inchon,* 114.

18. Clark, *The Secrets of Inchon,* 199.

19. Clark, *The Secrets of Inchon,* 189.

20. Clark, *The Secrets of Inchon,* 213–22.

21. Clark, *The Secrets of Inchon,* 214.

22. Clark, *The Secrets of Inchon,* 216.

23. Clark, *The Secrets of Inchon,* 239.

24. Clark, *The Secrets of Inchon,* 251.

25. Clark, *The Secrets of Inchon,* 254.

26. Clark, *The Secrets of Inchon,* 262.

27. Clark, *The Secrets of Inchon,* 301.

28. Clark, *The Secrets of Inchon,* 318.

29. *The Landing at Inch'on,* chapter 25, US Army History, https://www.history.army.mil/books/korea/20-2-1/sn25.htm.

30. Clark, *The Secrets of Inchon,* 325.

Afterword

1. Sverdrup and Munk, *Wind, Sea and Swell,* 12.

2. Bates and Fuller, *America's Weather Warriors,* 277.

3. Charles C. Bates obituary in Sahuaritasun.com Sun newspaper, 17 July 2016, https://www.sahuaritasun.com/obituaries/charles-c-bates/article_ca801f2a-4ab6-11e6-b6ae-e32cf622219e.html.

BIBLIOGRAPHY

Atkinson, Rick. *The Guns at Last Light: The War in Western Europe, 1944–1945*. New York: Henry Holt, 2013.

Alexander, Joseph H., Col., USMC (Ret.). "Across the Reef: The Marine Assault of Tarawa." National Parks Service, Marines in World War II Commemorative Series. https://www.nps.gov/parkhistory/online_books/npswapa/extContent/usmc/pcn-190-003120-00/sec8.htm.

Appleman, Roy E. *South to the Naktong, North to the Yalu (June–November 1950)*. Washington, DC: Center of Military History United States Army, 1992.

Australian Government, Bureau of Meteorology. "Dodge Tide." http://www.bom.gov.au/oceanography/projects/ntc/dodge/dodge.shtml.

Bates, Charles C. "Sea, Swell and Surf Forecasting for D-Day and Beyond: The Anglo-American Effort, 1943–1945." Unpublished manuscript, 2010.

———. "Sea, Swell, and Surf Forecasting for Operation Overlord." In *Some Meteorological Aspects of the D-Day Invasion of Europe, 6 June 1944,* edited by R. H. Shaw and W. Innes, 30–38. Boston: American Meteorological Society, 1986. Paper originally presented during symposium proceedings, Fort Ord, CA, 19 May 1984.

———. "Utilization of Wave Forecasting in the Invasions of Normandy, Burma, and Japan." *Annals New York Academy of Sciences*.51, no. 3 (May 1949):545–69.

Bates, Charles C., and J. F. Fuller. *America's Weather Warriors, 1814–1985*. College Station: Texas A&M University Press, 1986.

Bauer, K. Jack. *Surfboats and Horse Marines: U.S. Naval Operations in the Mexican War, 1846–48* Annapolis, MD: US Naval Institute, 1969.

Beaching the L.S.T. LST training video, Atlantic Fleet Amphibious Force Landing Craft Group, Photographic Science Laboratory, Naval Support Facility Anacostia, Washington, DC. YouTube. https://www.youtube.com/watch?v=9eCCgRCzN1Y.

"Beaufort Wind Scale." Stormfax Weather Almanac. http://www.stormfax.com/beaufort.htm.

Blair, Clay. *The Forgotten War: America in Korea, 1950–1953.* New York: Time Books, 1987.

Bundgaard, Robert C. "Forecasts Leading to the Postponement of D-Day." In *Some Meteorological Aspects of the D-Day Invasion of Europe, 6 June 1944,* edited by R. H. Shaw and W. Innes, 13–22. Boston: American Meteorological Society. 1984. Paper originally presented during symposium proceedings, Fort Ord, CA, 19 May 1984.

"Burntcoat Head Park." Atlas Obscura. https://www.atlasobscura.com/places/burntcoat-head-park.

Butikov, E. I. "A Dynamical Picture of the Oceanic Tides." *American Journal of Physics* 70, no. 10 (September 2002): 1001–11.

Casey, Hugh J., Major General, Chief Engineer. *Engineers of the Southwest Pacific, 1941–1945.* Vol. 4, *Amphibian Engineer Operations.* Washington, DC: Government Printing Office. 1959.

——. *Engineers of the Southwest Pacific, 1941–1945.* Vol. 8, *Critique.* Washington, DC: Government Printing Office. 1959.

Chief of Naval Operations, Aerology Section. "The Occupation of the Gilbert Islands." NAVAER 50-30T-4. Washington, DC, October 1944. http://www.ibiblio.org/hyperwar/USN/rep/NAVAER/Gilberts/index.html.

Clark, Eugene Franklin. *The Secrets of Inchon: The Untold Story of the Most Daring Covert Mission of the Korean War.* New York: G. P. Putman's Sons, 2002.

"Command History: Seventh Amphibious Force." Part I. HyperWar: A Hypertext History of the Second World War. http://www.ibiblio.org/hyperwar/USN/Admin-Hist/OA/419-7thAmphib/7thAmphibs-1.html.

Crowl, Philip A., and Edmund G. Love. *U.S. Army in World War II, The War in the Pacific: Seizure of the Gilberts and Marshalls.* Chapter 10. http://www.ibiblio.org/hyperwar/USA/USA-P-Gilberts/USA-P-Gilberts-10.html.

Crowell, John C. "Sea, Swell, and Surf Forecasting Methods Employed for the Invasion of Normandy, July 1944." Master's thesis, University of California at Los Angeles, 1946.

——. *Surf Forecasting for Invasions during World War II.* Santa Cruz, CA: Marty Magic Books, 2010.

Day, Deborah. "Walter Heinrich Munk Biography." Scripps Institution of Oceanography Archives, 2005. http://scilib.ucsd.edu/sio/biogr/Munk_Biogr.pdf

Doodson, A. T., and H. D. Warburg. *Admiralty Manual of Tides.* 1941; rpt., London: Her Majesty's Stationery Office, 1952.

Douglas, C. K. M. "Forecasting for the D-Day Landings." *Meteorological Magazine* 81, no. 960 (June 1952): 161–92.

Dyer, George C. *The Amphibians Came to Conquer.* Chapter 18, "That Real Toughie—Tarawa." HyperWar: A Hypertext History of the Second World War. http://www.ibiblio.org/hyperwar/USN/ACTC/actc-18.html.

Ellis, Earl H., USMC. *Advanced Base Operations in Micronesia.* HyperWar: A Hypertext History of the Second World War. http://www.ibiblio.org/hyperwar/USMC/ref/AdvBaseOps/index.html.

Garland, Albert N., and Howard M. Smyth. *The Mediterranean Theater of Operations: Sicily and the Surrender of Italy.* United States Army in World War II. Center for Military History. Washington, DC: Government Printing Office, 1965.

Geib, Claudia. "How a Father and Son Helped Create Weather Forecasting as We Know It." Atlas Obscura, 22 November 2016. https://www.atlasobscura.com/articles/how-a-father-and-son-helped-create-weather-forecasting-as-we-know-it.

Glines, C. V. "Airmail Service: It Began with Army Air Service Pilots." HistoryNet, 12 June 2006. http://www.historynet.com/airmail-service-it-began-with-army-air-service-pilots.htm. Originally published in *Aviation History,* May 1994.

Hammond, Major Jeffrey L., USMC. *Marine Corps Interwar Period Innovation and Implications for the Upcoming Post Operation Enduring Freedom Period.*_ http://www.dtic.mil/dtic/tr/fulltext/u2/a583991.pdf.

Harrison, Gordon A. *United States Army in World War II: The European Theater of Operations; Cross-Channel Attack.* Center for Military History, CMH Pub 7-4. Washington, DC: Government Printing Office, 1951.

Headquarters, Department of the Army. *Army Forces in Amphibious Operations (The Army Landing Force).* Field Manual 31-12, 28 March 1961. http://www.enlistment.us/field-manuals/fm-31-12-army-forces-in-amphibious-operations.shtml.

———. *Army Water Transport Operations.* Field Manual 55-50, 30 September 1993. http://www.enlistment.us/field-manuals/fm-55-50-army-water-transport-operations.shtml

Headquarters, European Theater of Operations, US Army. "Acknowledgement of Service Being Provided by 21st Weather Squadron." Ref. A/12. 14 August 1944. Copy in Reid Memoirs.

Heinl, Robert Debs, Jr. *Victory at High Tide: The Inchon-Seoul Campaign.* Annapolis, MD: Nautical and Aviation Publishing Company of America, 1980.

Higgins, Trumbull. *Korea and the Fall of MacArthur: A Précis in Limited War.* New York: Oxford University Press, 1960.

Isely, Jeter A., and Philip A. Crowl. *The U.S. Marines and Amphibious War: Its Theory, and Its Practice in the Pacific.* Princeton, NJ: Princeton University Press, 1951.

Kagan, Neil, and Stephen G. Hyslop. *Eyewitness to World War II: Unforgettable Stories and Photographs from History's Greatest Conflict.* Washington, DC: National Geographic Society. 2012.

Kamel, A. M., and D. D. Davidson. *Hydraulic Characteristics of Mobile Breakwaters Composed of Tires or Spheres.* Technical Report H-68-2. Vicksburg, MS: US Army Engineer Waterways Experiment Station, 1968.

Kinsman, Blair. *Wind Waves: Their Generation and Propagation on the Ocean Surface.* Englewood Cliffs, NJ.: Prentice Hall, 1965.

Krick, Irving P. "Role of Caltech Meteorology in the D-Day Forecast." In *Some Meteorological Aspects of the D-Day Invasion of Europe, 6 June 1944,* edited by R. H. Shaw and W. Innes, 24–26. Boston: American Meteorological Society, 1986. Paper originally presented during symposium proceedings, Fort Ord, CA, 19 May 1984.

"Lieutenant General Donald N. Yates." Air Weather Association. http://www.airweaassn.org/Library/People/LIEUTENANT%20GENERAL%20DONALD%20N_%20YATES.htm.

"Lieutenant General Thomas S. Moorman Sr." Air Weather Association. http://www.airweaassn.org/Library/People/LIEUTENANT%20GENERAL%20THOMAS%20S_%20MOORMAN%20JR.htm.

The Man Who Never Was: The True Story of Glyndwr Michael. http://www.themanwhoneverwas.com/.

Millett, Allan R. "Assault from the Sea: The Development of Amphibious Warfare between the Wars—The American, British, and Japanese Experiences." In *Military Innovations in the Interwar Period,* ed. Williamson Murray and Allan R. Millett, 50–95. Cambridge: Cambridge University Press, 1996.

Munk, Walter, and Deborah Day. "Harald U. Sverdrup and the War Years." *Oceanography* 15, no. 4 (2002): 7–29.

Neumann, Gerhard, and Willard J. Pierson Jr. *Principles of Physical Oceanography.* Englewood Cliffs, NJ: Prentice Hall, 1966.

Nierenberg, William A. "Harald Ulrik Sverdrup, 1888–1957." National Academy of Sciences, Biographical Memoirs. http://www.nasonline.org/publications/biographical-memoirs/memoir-pdfs/sverdrup-harald.pdf.

NOAA. "NOAA Celebrates 200 Years of Science, Service, and Stewardship." https://celebrating200years.noaa.gov/foundations/welcome.html

———. *Our Restless Tides.* https://tidesandcurrents.noaa.gov/restles3.html.

Pace, Eric. "Donald Pritchard, 76, Professor at SUNY and an Oceanographer." *New York Times,* 26 April 1999. http://www.nytimes.com/1999/04/26/nyregion/donald-pritchard-76-professor-at-suny-and-an-oceanographer.html.

"Operation Mincemeat—As Told by Ewen Montagu: The Operation Orders." In *The Man Who Never Was: The True Story of Glyndwr Michael.* http://www.themanwhoneverwas.com/operationmincemeat.html.

Parker, Bruce. "The Tide Predictions for D-Day." *Physics Today* 64, no. 9 (01 September 2011), https://doi.org/10.1063/PT.3.1257.

Paulus, Robert D. "Pack Mules and Surfboats: Logistics in the Mexican War." *Army Logistician* 29, no. 6 (1997): 34–40.

Quinn, Mary-Louise. *The History of the Beach Erosion Board, U.S. Army Corps of Engineers, 1930–63.* Miscellaneous Report 77-9. Fort Belvoir, VA: US Army Corps of Engineers, 1977.

Rees, David. *Korea: The Limited War.* Baltimore: Penguin Books, 1964.

Reid, Jan. "Texas Primer: The Blue Norther." *Texas Monthly,* November 1982, https://www.texasmonthly.com/the-culture/texas-primer-the-blue-norther/.

Reid, Robert O. "Robert Osborne Reid Memoirs." Unpublished manuscript. Vol. 1, "The Formative Years: 1921–1951." 2008.

Rosenthal, Hans E. "Northers of the Gulf of Mexico and Central American Waters." *Mariners Weather Log,* November 1965.

Ross, John. *The Forecast for D-Day and the Weatherman behind Ike's Greatest Gamble.* Guilford, CT: Lyons. 2014.

Ruppenthal, Roland G. *Logistical Support of the Armies.* Vol. 2, *September 1943–May 1945.* United States Army in World War II, The European Theater of Operations. Washington, DC: Center of Military History, United States Army, 1995.

Schaefer, Chris. *Bataan Diary: An American Family in World War II, 1941–1945.* Houston: Riverview, 2004.

Simmons, Edwin Howard. "Amphibious Warfare." In *The Oxford Companion to American Military History,* edited by John Whiteclay Chambers II et al., 31–32. Oxford: Oxford University Press, 1999.

Stagg, J. M. *Forecast for Overlord, June 6, 1944.* New York: W. W. Norton, 1971.

Stevenson, Fanny Van de Grift. *The Cruise of the* Janet Nichol *among the South Sea Islands.* New York: Charles Scribner's Son, 1914.

Stevenson, Robert Louis. *In the South Seas.* Web edition published by eBooks@Adelaide, University of Adelaide, South Australia, 2010. http://ebooks.adelaide.edu.au/s/stevenson/robert_louis/s848so/index.html.

Stockman, Captain James R., USMC, *The Battle for Tarawa,* USMC Historical Monograph. http://www.ibiblio.org/hyperwar/USMC/USMC-M-Tarawa/.

Stout, Dorothy, L. "John C. Crowell." *Geotimes,* June 1999, 18–24.

Strahan, Jerry E. *Andrew Jackson Higgins and the Boats That Won World War II.* Baton Rouge: Louisiana State University Press, 1994.

Sverdrup, H. U., W. M. Johnson, and R. H. Flemming. *The Oceans: Their Physics, Chemistry and General Biology.* Englewood Cliffs, NJ.: Prentice-Hall, 1942.

Sverdrup, H. U., and W. H. Munk. *Wind, Sea and Swell: Theory of Relations for Forecasting.* U.S. Navy Department, Hydrographic Office, pub. no. 601, 1947.
25th US Naval Construction Regiment. "Action Report for June 6 to 22, 1944." http://www.seabeecook.com/history/25th_ncr/action_report/action_report_body.htm.
Wukovits, John. *One Square Mile of Hell: The Battle for Tarawa.* New York: New American Library Caliber, 2006.

INDEX